Enlistment Decisions in the 1990s

Evidence From Individual-Level Data

M. Rebecca Kilburn
Jacob A. Klerman

Prepared for the
Office of the Secretary of Defense • U.S. Army

National Defense Research Institute • Arroyo Center

RAND

For more information on the RAND Arroyo Center, contact the Director of Operations, (310) 393-0411, extension 6500, or visit the Arroyo Center's Web site at http://www.rand.org/organization/ard/

PREFACE

This report estimates a model of individual enlistment decisions using the National Educational Longitudinal Study (NELS) sample. The model incorporates our imputations of AFQT scores for NELS respondents described in an earlier phase of the project entitled "Recruiting Policy and Resources" and reported in MR-818-OSD/A, *Estimating AFQT Scores for NELS Respondents* (Kilburn, Hanser, and Klerman, 1998). Other results from the project are presented in additional RAND documents: MR-549-A/OSD, *Recent Recruiting Trends and Their Implications: Preliminary Analysis and Recommendations* (Asch and Orvis, 1994); MR-677-A/OSD, *Military Recruiting Outlook: Recent Trends in Enlistment Propensity and Conversion of Potential Enlisted Supply* (Orvis, Sastry, and McDonald, 1996); MR-847-OSD/A, *Recent Recruiting Trends and Their Implications for Models of Enlistment Supply* (Murray and McDonald, 1998); and MR-845-OSD/A, *Encouraging Recruiter Achievement: A Recent History of Recruiter Incentive Programs* (Oken and Asch, 1997). This report provides insights for individuals setting recruiting policy. It will also be of interest to those concerned with the broader post-secondary activity choices of youths.

This research was conducted for the Under Secretary of Defense for Personnel and Readiness and for the Deputy Chief of Staff for Personnel, U.S. Army. The project was executed jointly by the Forces and Resources Policy Center of RAND's National Defense Research Institute (NDRI) and the Manpower and Training Program of the RAND Arroyo Center. The Arroyo Center and NDRI are both federally funded research and development centers, the first sponsored by the United States Army and the second sponsored by the Office of the Secretary of Defense, the Joint Staff, the unified commands, and the defense agencies.

CONTENTS

Preface iii

Tables vii

Summary ix

Acknowledgments xix

Chapter One
INTRODUCTION 1

Chapter Two
MODEL OF INDIVIDUAL ENLISTMENT DECISIONS 5
The Random Utility Framework 5
Hosek and Peterson Specification 8
Determinants of Individual Decisions: Supply Factors . 9
Determinants of Military Behavior: Demand Factors . . 12

Chapter Three
CHANGES IN THE RECRUITING ENVIRONMENT 15
Supply Factors 16
Shrinking Youth Cohort 16
Greater Share of Minorities and Immigrants in Youth Cohorts 16
Increase in College Attendance 18
Youth Aptitudes Have Risen 19
Increase in Military Pay and the Decline in Real Wages of High School Graduates 20
Desert Storm and Rise in Deployments 20
Demand Factors 21
Smaller Recruiting Requirements 21

Rising Recruit Quality 22
Expansion of Women's Roles in Military 22
Cutbacks in Advertising 23
Reduction in Number of Recruiters 23
Changes in Recruiter Management 24
Altering Enlistment Incentive Programs 24
Higher Crime Rate Among Youth 25
Innovations in the Specification 25
Estimating a Three-Choice Model of Enlistment 28

Chapter Four
NELS DATA 31

Chapter Five
RESULTS 37
Replicating Hosek and Peterson 37
New Bivariate Specification 55
Estimating the Trivariate Choice Model 60

Chapter Six
CONCLUSIONS 65

Appendix

A. SUPPLEMENTARY TABLES 69

B. TESTING FOR DIFFERENCES FROM
HOSEK AND PETERSON RESULTS 79

References 83

TABLES

S.1. Characteristics Significantly Affecting Enlistment Probability for Seniors and Graduates xi
S.2. New Variables Added to the Enlistment Model for Seniors and Graduates xiii
S.3. Characteristics Significantly Affecting Probability of Choosing College or Work Relative to Enlisting xvi
3.1. Percent of Seniors and Graduates Enlisting as of 1994, by Estimated AFQT Category 28
5.1. Logistic Regression Estimates of Enlistment Probability Using Hosek and Peterson Specification, for Seniors and Graduates 39
5.2. Mean Predicted Probability of Enlisting for Seniors and Graduates, by Enlistment Status 41
5.3. Comparing NELS and Hosek and Peterson (H&P) Coefficient Estimates, Seniors Model 43
5.4. Comparing NELS and Hosek and Peterson (H&P) Coefficient Estimates, Graduates Model 45
5.5. Comparing NELS and Hosek and Peterson (H&P) Regressor Means, Seniors Model 48
5.6. Comparing NELS and Hosek and Peterson (H&P) Regressor Means, Graduates Model 50
5.7. Logistic Regression Estimates of Enlistment Probability Using New Specification, for Seniors and Graduates 57
5.8. Mean Predicted Probability of Enlisting for Seniors and Graduates, by Enlistment Status 59
5.9. Percent Making Each Choice in the Trivariate Model . 60

5.10. Change in Probability of Making Each Choice for One Unit Change in Regressor 61
5.11. Mean Predicted Probability of Making Each Choice, by Actual Choice 64
A.1. Variable Definitions 69
A.2. Means and Standard Deviations of Variables 74
A.3. Multinomial Logit Estimates for Trivariate Model 76

SUMMARY

The current recruiting environment is marked by recruiting targets higher than those in the drawdown period, increasing competition for resources, and reports that recruiters are having more difficulty in meeting their goals. Models that specify persons with different characteristics and predict the probability that they will enlist could thus help allocate current recruiting resources to target the most likely prospects. However, the most recent individual-level models of enlistment were estimated using data from 1980. Since then, many trends and events suggest that the enlistment likelihoods among different types of individuals may have changed. These trends include an increase in college attendance, shrinking youth cohorts, rising youth aptitudes, and an increase in the number and scope of deployments. In this report, we take several approaches to updating the principal economic model of enlistment decisionmaking with data from 1992 and 1994.

MODEL AND DATA

Our model of enlistment decisionmaking is grounded in the economic theory of individual choice. The data are drawn from large, ongoing surveys that follow individual youths over a period of years to generate information relevant to policymaking and the social sciences. Relevant variables include race and ethnicity, aptitude, plans for marriage and education, family income, and various parental characteristics. We drew other important variables describing labor markets from the Census Bureau. Finally, we obtained enlistment information from survey participants either from the survey itself or from other sources.

Statistical methods such as logistic regression then permit the specification of an equation in which the probability of enlistment is set equal to the sum of a series of terms. Each term is specific to a particular characteristic and is the product of a coefficient and the value of that characteristic (e.g., score on an aptitude test). These coefficients express the degree to which changes in the value of the characteristic influence the enlistment probability, and in which direction.

It has been found that youths can be separated into broad groupings—e.g., high school seniors versus high school graduates, men versus women—whose enlistment behavior responds sufficiently differently to changes in other characteristics to warrant the creation of separate models. In updating these models, we use data on high school seniors from interviews conducted in 1992 for the National Educational Longitudinal Study (NELS). We use data on high school graduates from NELS interviews conducted in 1994. We supplement these with 1990 Census data. From this we learn the factors associated with decisions by young men to enlist during their senior year (that is, sign a contract promising to enter the service after graduation). And we learn the factors associated with decisions to enlist by high school graduates at approximately age 20 who are not then enrolled in college. We estimate models for young men only. The number of enlisting women in NELS is too small to permit estimating coefficients for them with any level of confidence.

We take three approaches to updating the models:

- We reestimate the earlier model, maintaining the original specification of variables and two-way decision to enlist, but using the new data.
- We change the set of variables in an effort to derive a model that is more useful.
- We change the form of the model from a representation of a two-way decision, to enlist or not, to a three-way representation of a decision to enlist, attend college, or join the workforce.

USING THE SAME VARIABLES

Our replication of the Hosek and Peterson models (1985, 1990), which were estimated using data from the early 1980s, generally finds that similar variables raised the likelihood of enlistment in the NELS data from the early 1990s. In other words, despite the many changes that took place between 1980 and 1992, the same characteristics were associated with enlistment decisions in the two periods. Variables that might be expected to be associated with college attendance were the strongest influences on high school seniors' enlistment decisions, whereas variables associated with job opportunities most strongly affected graduates' decisions, as shown in Table S.1. This table lists the characteristics that exhibited a statistically significant influence on the enlistment decisions for seniors and graduates in the NELS data. The plus or minus indicates whether the characteristic is associated with a higher or lower probability of enlisting, respectively.

Table S.1

Characteristics Significantly Affecting Enlistment Probability for Seniors and Graduates

Seniors	Graduates
Hispanic (–)	Family income (–)
AFQT score (–)	Lives at home (–)
Mother's education (–)	Weekly hours worked currently (+)
Family income (–)	Not employed (+)
Number of siblings (+)	Weekly hours worked, previous job (+)
Low wage (–)	Months not employed (–)
Months not employed (–)	Not employed last 12 months (+)
Unemployment rate (+)	Recruiter density (–)
Unemployed times months not employed (+)	Ever married (+)
Plans to marry within 5 years (+)	Has children (–)

NOTES: Plus sign indicates greater enlistment probability with greater variable value (or "yes" where variable is dichotomous); minus indicates lower enlistment probability with greater variable value (or "yes"). Not all factors in the model are shown here. AFQT = Armed Forces Qualification Test, an aptitude measure.

For seniors, we find that AFQT score, mother's schooling, family income, number of siblings, marital plans, and only a couple of work-related variables are important determinants of enlistment choice. For graduates, family income, number of siblings, and marital plans were still related to enlistment probability, but a great number of the work-related variables were also important, including wage-related variables, employment status, duration of unemployment, and other variables. Our measure of the number of recruiters per potential recruit, *Recruiter density,* is only statistically significant for graduates but yields an unexpected negative coefficient. This result may be due to the fact that recruiting assignments made by the services are related to factors such as the ease or difficulty of recruiting in an area.

Note that a number of variables have no significant relation to enlistment behavior. In particular, previous studies have found higher enlistment rates for blacks, which is not the case here. Other RAND research has shown a drop in the interest expressed by black youths in joining the military (Orvis, Sastry, and McDonald, 1996). Overall, about a quarter of the senior coefficients and a third of the graduate coefficients differ significantly from those in the earlier model. Hence, despite the numerous changes in the recruiting environment that took place between 1980, when previous estimates were generated, and 1992–1994, the years our data were collected, we generally find that the same characteristics predict enlistment in the two periods.

ADDITIONAL VARIABLES

We also sought to take into account social trends potentially affecting enlistment decisions and to improve the model through the addition or omission of variables. Three social trends we sought to reflect in the model were growth in immigration, drug use, and crime. Immigrants have steadily increased their share of the U.S. population since Hosek and Peterson estimated an enlistment model. We did not have data for immigration status, but as a proxy we used whether the person's first language was something other than English. Drug use also grew between 1980 and the early 1990s, and we included a variable indicating whether the respondent had ever used marijuana. Crime was also higher, particularly among youths, in the early 1990s than in 1980. Crime is of interest because the military would like to avoid enlisting youths with an arrest record. Our crime-related variable indicated whether the respondent or one of his friends had ever been arrested. We also added several other variables that might improve the model, including whether the youth had a parent in the military and the average cost of in-state college tuition. We also reformulated the aptitude variables to allow for the possibility that enlistment probabilities might peak in the center of the aptitude distribution instead of at either end.

Several of these changes did yield new insights about enlistment probabilities, as shown in Table S.2. For seniors, the variable indicating that English is not a youth's first language substantially lowered the probability of enlistment. The diversity of the immigrant population and the divergence in patterns of college attendance among them is great: some Asian immigrant groups have much higher college attendance rates than average, while other immigrant groups, such as those from Central America, have much lower college attendance rates than average. Given this diversity, our finding on the effect of English being a second language warrants more exploration. That is, it may be premature to direct recruiting efforts away from immigrant groups when some immigrant groups may yield high numbers of recruits and others few.

Table S.2

New Variables Added to the Enlistment Model for Seniors and Graduates

Variable	Seniors	Graduates
Parent in military	Insignificant	(+)
English not first language	(–)	Insignificant
Uses marijuana	Insignificant	Insignificant
Respondent or friend has been arrested	Insignificant	(+)
Average in-state college tuition	Insignificant	Insignificant

NOTES: Plus sign indicates greater enlistment probability with greater variable value (or "yes" where variable is dichotomous); minus indicates lower enlistment probability with greater variable value (or "yes"). Insignificant means the variable was not statistically significant. Not all factors in the model are shown here.

For graduates, a new variable that significantly raised enlistment probability is having a parent in the military. This finding suggests potential for recruiting through veterans' organizations or other avenues for targeting youths with currently or formerly enlisted parents. Variables indicating marijuana use yielded some ambiguous but suggestive results: youths who did not answer questions about their marijuana use, which may be correlated with use, were substantially less likely to enlist than others.

A THREE-CHOICE MODEL OF ENLISTMENT

The models just described treat the enlistment decision as a "yes" or "no"—a youth decides to enlist or not to. Another way of characterizing the decision to enlist is to treat the decision as between enlisting and multiple civilian alternative activities. Hence, we also estimated an individual enlistment model in which the enlistment decision is a three-way choice among enlisting in the armed forces, going to college, or working. We estimated this model with a sample that pooled the senior and graduate groups.

The most important reason for estimating the three-choice model is that it shows which activities youth are likely to choose if they do not enlist in the military. This is important for designing recruiting incentives because it allows the military to tailor the incentives to draw recruits away from the next-best alternative. For example, if college attendance is the next-best alternative, recruiting incentives might want to stress educational benefits or on-the-job training. But if civilian employment is the next-best alternative, recruiting incentives might focus on job security, wage comparability, or benefits.

Many of the estimates of the trivariate model imply a high degree of substitutability between the college/military choice for seniors. This result implies that competition for recruits who are seniors derives largely from higher-education opportunities and hence recruiting resources should be directed in a way that recognizes college as an important alternative activity. The results also point to civilian employment as the most important source of competition for recruits who are graduates.

A second advantage of the three-choice model is that the results provide more insights into the role of particular variables in the enlistment decision. Variables included in individual enlistment models are typically included on the basis that they represent the advantage or disadvantage to the individual of enlisting *relative* to some other specific alternative. By comparing enlistment to attending college and working, the results show much more clearly whether a characteristic that predicts enlistment does so because it makes recruiting more attractive relative to college or relative to work.

The results of the three-choice model are also more plausible than those of the two-choice model. First, the same set of variables considered in the previous model (Table S.2) yields a larger number of significant relations. This is because some variables are positively associated with a decision to attend college instead of enlisting and negatively associated with a decision to work instead of enlisting (or vice versa). These opposite associations could cancel each other out when estimating a simple enlist-or-not model. Second, variables were chosen for previous enlistment models because of their hypothesized relation with decisions to attend college instead of enlisting or take a job in the civilian sector instead of enlisting. With this new specification, we can now test those hypotheses directly.

Variables that are included in the model primarily because they are believed to influence the expected returns to education—such as AFQT score, age when a senior, and mother's education—lower enlistment rates because they raise the likelihood of attending college rather than by operating through the work/other choice (see Table S.3). These findings suggest that youth considering college may respond well to recruiting incentives associated with attending college, such as college benefit programs.

As shown in Table S.3, when contrasting the decision to enlist or attend college, there was no significant difference between the probability of choosing one or the other for youths in AFQT CAT I through AFQT CAT IIIA, but youths with lower test scores were less likely to choose to attend college. Hence the biggest difference between college attendees and high-quality recruits may not be test scores but rather being from a disadvantaged background. We found variables associated with the availability of resources to pay for college, mother's education, and early marriage and childbearing to be strong predictors of which high-quality youth attended college and which enlisted. For individuals who have chosen not to attend college and

Table S.3

Characteristics Significantly Affecting Probability of Choosing College or Work Relative to Enlisting

College	Work
Black (–)	Predicted AFQT CAT I (–)
Predicted AFQT CAT IIIB (–)	Predicted AFQT CAT II (–)
Predicted AFQT CAT IV (–)	Predicted AFQT CAT IIIB (+)
Predicted AFQT CAT V (–)[a]	Predicted AFQT CAT IV (+)[a]
GED (–)	Predicted AFQT CAT V (+)[a]
Mother's education: less than high school (–)	Mother worked (–)
Mother's education: college degree (+)	Very low family income (+)
Mother's education: postcollegiate (+)	Number of siblings (+)
Family income (+)	Percent of population black (–)
Very low family income (+)	Per-capita personal income (–)
Number of siblings (–)	Recruiter density (+)
Unemployment rate (–)	Ever been married (+)
Per-capita personal income (+)	Has children (+)
Recruiter density (–)	Parent in the military (+)
Expects more education (+)	Uses marijuana (+)
Plans to marry within 5 years (–)	
Plans never to marry (–)	
Ever been married (–)	
Has children (–)	
Parent in the military (–)	
English not first language (+)	
Youth or friend has been arrested (–)	

NOTES: Plus sign indicates greater probability of choosing college or work with greater variable value (or "yes" where variable is dichotomous); minus indicates lower probability with greater variable value (or "yes"). Not all factors in the model are shown here. AFQT = Armed Forces Qualification Test. AFQT CAT indicates the percentile range in which the person scored, with CAT I being highest and CAT V being lowest.

[a]Note that these are *predicted* scores as of the person's senior year in high school, not the actual score used for military entrance.

are considering working or enlisting, the middle to lower part of the AFQT distribution might be more fertile ground for recruiting efforts. We also observe that family socioeconomic status has less impact on recruiting decisions for this group, although marriage and fertility continue to have a strong influence.

Our measure of recruiter density appears to show that more recruiters per potential recruit is effective in attracting recruits away from college, but is less effective in competing with work. Among the new variables we have been examining, we find that having a parent in the military, English not being the person's first language, whether the youth or a friend had been arrested, and marijuana use are also important predictors of enlistment in the three-choice model.

CONCLUSION

Despite the numerous changes that have occurred between the early 1980s and 1992–1994, our estimates of individual enlistment decision models generally find that the same variables were important predictors of enlistment in the two periods. To try to capture some of the changes since the earlier models were estimated, we also modified the specifications used in those studies. We found a few new variables that were predictive of enlistment, namely a proxy for immigrant status for seniors, and having a parent in the military or having been arrested (or having a friend who had been arrested) for graduates. Given that these variables are really just proxies for other underlying concepts, these findings warrant further exploration.

We also furthered understanding about the competition military recruitment faces with results from a three-choice model, which compares the probability of choosing to enlist in the military, enter college, or work after high school. This model suggests a high degree of substitutability between college and the military for high-quality youth, and between work and the military for other young men. Hence, to attract high-quality youth, recruiting incentives should focus on attracting college-bound youth rather than on incentives oriented toward the labor market. For graduates who have clearly chosen not to attend college, the opposite is true.

ACKNOWLEDGMENTS

This project received helpful input from many individuals. First, we thank all the individuals who helped us with the test score imputation that was conducted as the first step of this study and is reported in MR-818-OSD/A, *Estimating AFQT Scores for NELS Respondents* (Kilburn, Hanser, and Klerman, 1998). This includes our reviewers, John Adams and Brian Stecher, and our co-author, Larry Hanser. Second, we thank Laurie McDonald, who provided strong programming support for the entire project, Nora Wolverton, who assisted in the preparation of the manuscripts, and James Chiesa for help on the summary. Third, we thank project leaders Beth Asch and Bruce Orvis for their support and guidance. Fourth, we thank our sponsors, especially Dr. Steve Sellman, Director of Accession Policy, for valuable input. Finally, we thank James Hosek and John Warner for their helpful reviews.

Chapter One

INTRODUCTION

After declining continuously since 1989, total Department of Defense (DoD) accessions increased for the first time in 1996. In 1997, total accessions rose again, posting a gain of nearly 5 percent. During the drawdown, however, recruiting resources had been cut dramatically and continued to remain at relatively low levels until the last half of this decade. Coupled with reports of declining youth propensity to enlist, a strong civilian labor market, and recruiters having difficulty meeting goals (see discussion in Orvis, Sastry, and McDonald, 1996) this raised the military's concerns about its ability to successfully recruit at higher target levels. This document is part of a larger project that examines recent trends in recruiting and their implications for DoD policy and expectations of near-term recruiting success.

The overall project uses a combination of complementary approaches to study recruiting. One approach uses recent waves of survey results from the Youth Attitude Tracking Study (YATS) to study how youths' perceptions explain recent recruiting trends and the implications of trends in these perceptions for recruiting outcomes (see Orvis, Sastry, and McDonald, 1996).

The second approach the project uses to understand recent recruiting trends explores the aggregate relationships between recruiting outcomes and a number of macro-level variables including recruiting policy, civilian opportunities, and youth population trends. Murray and McDonald (1998) indicates how the number of accessions in a geographic area changes when certain national-level and local-area variables change. These variables include unemployment rates and youth population, among others. Hence, the aggregate-level models suggest how policy variables and future macro-level trends such as unemployment rates or test score trends are likely to influence DoD's ability to attract recruits.

This report uses a third approach to examine recent recruiting trends. We investigate the determinants of individual-level enlistment decisions. The research investigating individual enlistment decisions answers questions about the characteristics of individuals who enlist and what factors influence the choices of those who do and do not enlist. These characteristics include race and ethnicity, marital status, and family background variables. Other factors that may influence individual enlistment decisions are local labor-market conditions and local higher education costs and opportunities. The individual-level models indicate which types of individuals are most likely to enlist so recruiters can target these individuals, helping DoD obtain the required number of recruits at the lowest possible outlay of recruiting resources.

The last DoD-sponsored studies using an economic model of individual enlistment decisions were Hosek and Peterson (1985), which analyzed the enlistment decisions of young men, and Hosek and Peterson (1990), which compared the enlistment decisions of young men and women. These studies used the 1980 wave of the National Longitudinal Survey of Youth (NLSY) and the 1979 DoD Survey of Personnel Entering Military Service. They show that among those eligible to enlist, the decisionmaking process differs according to whether the youth is a high school senior or a graduate, and whether the youth is male or female.

The data used in the Hosek and Peterson studies are now nearly two decades old. Since the late 1970s, numerous changes have taken place that may have influenced individual enlistment probabilities. Among them are: the youth cohort is slightly smaller, the total number of recruits needed has fallen sharply, a larger share of youths are minorities and immigrants, youth aptitudes have risen, recruiter management has changed, a higher fraction of youths are attending college, the earnings of high school graduates have declined relative to college graduates, more recruits are female, the military experienced the drawdown, and we engaged in the first war since Vietnam.

This study estimates individual enlistment decision models using a more recent data set, the National Educational Longitudinal Study (NELS), which was fielded in 1992 and 1994. First, we replicate the Hosek and Peterson studies (1985, 1990) with the more recent NELS data. Our objective in conducting this replication is to identify the extent to which there were changes in the variables that predict individual enlistment choices between 1980 and 1992.

Next, we estimate additional models of enlistment decisions that expand upon the model of Hosek and Peterson. The primary innovation we make in our extension of this model is to explore the role of educational activities as alternatives to military enlistment. Although their theoretical discussion recognizes that enlistment involves an individual's choice among school, work, and military service, data limitations required Hosek and Peterson to estimate an enlistment model as a bivariate choice between enlistment and other activities. Their model includes variables that measure civilian labor-market conditions to capture the relative attractiveness of the other activities. Because of the increase in postsecondary school attendance—particularly for populations of youths with characteristics similar to military recruits—we include variables intended to capture the attractiveness of attending college instead of enlisting. We examine the viability of expanding the set of alternatives to military service in two ways. First, we estimate the bivariate model including additional schooling variables. Second, we estimate a trivariate model that not only includes the additional schooling variables, but also models the decision as being one between enlistment, civilian labor force participation, and enrolling in school. Other innovations to the model we explore are a slightly different way of including AFQT score in the model and including some additional covariates. This is motivated by the changes, enumerated earlier, in the recruiting environment since the time that Hosek and Peterson specified their model. Since AFQT is an important explanatory variable that is not included in the NELS data, an earlier phase of this project predicted AFQT scores for the NELS participants. These results are reported in detail in Kilburn et al. (1998).[1]

---

[1]The other report, Kilburn et al. (1998), also discusses some of the implications of the estimated test score patterns for recruiting policy.

This report has six chapters. Chapter Two outlines the decision model that provides the framework for studying individual enlistment decisions. Chapter Three describes changes such as civilian labor market and college enrollment patterns that would suggest the results reported in Hosek and Peterson might differ from those estimated with more contemporary data. Chapter Four describes the NELS. Chapter Five reports the results of our replication of the Hosek and Peterson specification and of our new three-choice specification of individual enlistment decisions. Chapter Six discusses our findings and their implications for recruiting policy.

Chapter Two

MODEL OF INDIVIDUAL ENLISTMENT DECISIONS

This chapter begins by presenting the economic model that serves as the theoretical framework for studies of individual enlistment decisions (see Hosek and Peterson, 1985, 1990; Kilburn, 1994; Kim et al., 1980; and Orvis and Gahart, 1985). Then we review the Hosek and Peterson specification of this framework—that is, what variables Hosek and Peterson included in their estimates. We review this model because the first objective of our report is to compare estimates of this specification of enlistment with the more recent NELS data to the Hosek and Peterson estimates, which use data from 1980, to see if the effects of variables changed during the 15-year period.

THE RANDOM UTILITY FRAMEWORK

Earlier economic models of individual enlistment decisions (Hosek and Peterson 1985, 1990; Gorman and Thomas, 1993; Kilburn, 1994; and Kim et al., 1980) are variants of the random utility model (McFadden, 1983). We outline the general random utility framework here, and later show how the Hosek and Peterson specification and our innovations are variants of this framework.

The individual who is deciding whether or not to enlist is eligible for military enlistment and can choose between enlistment and other activities such as college, employment, and working in the home. By assumption in the basic random utility framework, individuals choose the activity that yields the highest expected utility. An individual chooses to enlist in the military if the utility of enlisting is greater than the utility of the other alternatives, or

$$U_{im} > U_{ij} \text{ for } j=1,2\ldots J,$$

where U indicates utility, i represents the individual, m represents the military, and j represents nonmilitary alternatives.

This behavioral model is translated into a statistical model by expressing the likelihood that an individual makes the observed choice as a probability. The probability that an individual chooses to enlist over some other activity, j, is

$$\Pr\left(U_{im} > U_{ij}\right). \tag{1}$$

We let the approximate utility to individual i of alternative k be a function of characteristics of the individual x_i and a random error component ε_{ik} such that

$$U_{ik} = f_k\,(X_i) + \varepsilon_{ik}.$$

The X_i include characteristics such as the resources the individual's family has for funding educational investments, the person's AFQT score, and other characteristics that would be expected to alter the relative utility of the competing alternatives. Models of occupational or educational choice typically specify the utility of the alternatives as a function of potential wages, investment cost, earnings growth, or returns to investment (see Manski and Wise (1983), Willis and Rosen (1979), and Hosek and Peterson (1985), for example).

Rewriting equation (1) in terms of the observed characteristics of the individual and the error component produces the following expression for the probability that the individual chooses to enlist in the military, m:

$$\Pr\left([f_m\,(X_i) + \varepsilon_{im}] > [f_j(X_i) + \varepsilon_{ij}]\right) \text{ for all } j.$$

Assuming a linear form for the function $f_k(X_i)$ and an extreme value distribution for the error yields the multinomial logit model (McFadden, 1983):

$$\Pr(k = m) = \frac{e^{b'_m X_i}}{\sum_k e^{b'_k X_i}}.$$

This is the statistical model we estimate. It expresses the probability that the individual chooses choice m as a function of the characteristics of the individual and the attributes of the choices. The estimates of interest will be the coefficient values, β, and their significance levels.

The probability that each individual enlists rises as the coefficients on the individual characteristics and choice attributes are higher for enlistment than other alternatives. In terms of the equation above, this is equivalent to saying that the probability that an individual enlists is higher when $\beta_m > \beta_k$. This implies that having a particular characteristic raises the probability that the person enlists more than it raises the probability that the individual chooses one of the other alternatives.

Individual characteristics influence the probability that one person enlists relative to another person. One person will be more likely to enlist than another person if that individual has characteristics that tend to raise the utility of enlisting relative to other alternatives, or the probability of enlisting rises as the military alternative has attributes that raise the utility of enlisting relative to the other alternatives. For example, if a particular individual characteristic, say X_{i1}, raises the probability of enlisting more than that of choosing the alternatives ($\beta_{m1} > \beta_{k1}$), then people with higher levels of X_{i1} will be more likely to enlist than people with lower levels of X_{i1}.

Now we will put the research questions of this report in terms of the statistical model we just outlined. Our first objective is to compare estimates of the enlistment model Hosek and Peterson obtained from the 1980 NLSY data to estimates based on the more recent NELS data. In terms of the statistical model above, this involves two comparisons. First, we will compare the coefficient estimates we obtain for β using the NELS to the estimates Hosek and Peterson (1990) obtained. Second, we will compare the levels of the characteristics, X_i, in the NELS data to the levels in the NLSY data Hosek and Peterson used.

Our next objective is to explore the possibility that other variables should be included in the model. This means that we will add variables to those included by Hosek and Peterson and determine whether those variables are statistically significant in the estimates. In other words, we will add variables to the set Hosek and Peterson included in X_i and reestimate the model. As discussed in more detail below, we select variables to add based on theoretical considerations. Our final objective is to examine whether estimating the enlistment choice as compared to two alternative choices—college attendance and working—rather than one alternative choice—not enlisting—as in Hosek and Peterson provides additional insights. Like Hosek and Peterson (1990), we first estimate a model with only two choices, enlisting and not enlisting. Then we estimate a model with three choices—enlisting, attending college, and working—and compare these estimates to those from the two-choice model.

HOSEK AND PETERSON SPECIFICATION

Hosek and Peterson model the choice to enlist as a choice between two activities: enlisting and not enlisting. The individual's decision rule would then be to enter the military if the expected valuation of a military career exceeds the expected valuation of any other alternative. Our outline of the random utility model above stated that the model examined the choices of *eligible* individuals. Hosek and Peterson take individuals in their sample to be eligible if their AFQT score as measured in the NLSY is in the legally eligible range—that is, their score is above the 9th percentile (see Kilburn et al. (1998) for a more thorough discussion of eligibility based on test scores).

Now we discuss the variables that Hosek and Peterson (1985, 1990) included when they estimated the two-choice model. They base their specification on theories of career choice from the youth's perspective and theories of recruiter behavior from the DoD perspective. These papers model the decision to enlist of two segments of the recruiting market: high school seniors and nonstudent high school graduates—seniors and graduates, for short. The hypothesis behind segmenting the recruiting market is that the determinants of enlistment decisions vary by segment. Consistent with this, Hosek and Peterson (1985, 1990) in fact find that graduates' decisions respond more to work-related variables like wage rates, job tenure, labor force experience, employment status, and duration of joblessness. In contrast, the decisions of seniors are more responsive to variables related to education, including learning proficiency measures, the ability to finance additional education, and measures of parental influence. They also find that results differ according to whether the individual expects more education or not.

As noted by Dertouzos (1985), recruiting outcomes are not only the result of supply factors such as individual decisions about enlistment, but are also influenced by demand on the part of the military. Hence, in addition to factors that would affect individual decision-making, Hosek and Peterson incorporate into their model features that represent the military's demand for new recruits. These include factors like recruiter density and measures associated with enlistment standards such as whether the individual has a test score in the unacceptable range.

Here we briefly outline the factors that Hosek and Peterson include in their specification. First we review the variables that are related to recruit supply, and then we review the variables related to the demand for recruits from military recruiting and alternative activities like college and labor force participation.

Determinants of Individual Decisions: Supply Factors

Expected returns to education. The first factor in this category is learning proficiency. Individuals with higher learning proficiency would be expected to have higher returns to educational investments and would therefore be more likely to acquire more education after high school and less likely to enlist in the military. The variables we use to measure learning proficiency are estimated *Armed Forces Qualification Test (AFQT) score* and *age when a senior*. The AFQT is administered by the military to applicants to the enlisted force, and it is designed to measure successful completion of advanced military training, or trainability. We would expect that individuals with higher *AFQT scores* would have higher learning proficiencies. The higher is *age when a senior,* the lower we would expect learning proficiency to be. This is because individuals at a higher *age when a senior* have taken more years to complete high school than their younger counterparts.

Hosek and Peterson (1985) hypothesize that learning proficiency should be less important as a determinant for the enlistment choices of graduates than seniors. This is because graduates have revealed a low propensity to continue on in school, and since their likelihood of attending college should rise little with increases in learning proficiency, the measure of learning proficiency should have little impact on their enlistment decision.

A second factor in this category is educational expectations. To the extent that individuals expect more education, we would predict that they would be more likely to pursue more education and therefore less likely to enlist in the military. Variables that measure an individual's educational expectations include whether he or she *expects more education* and *mother's education*. The variable *mother's education* proxies for parents' education. This is because parents' education is highly correlated and the data are more likely to include mother's education than father's education, since fathers are absent from the household more often. We expect that youths whose parents have more education will themselves be more likely to obtain more education. This is because the cost of obtaining information about college application procedures, college alternatives, and careers for college graduates ought to be lower for individuals whose parents attended college and because more-educated parents may have a taste for higher education.

Hosek and Peterson (1985) indicate that educational expectations may have different effects on the likelihood of enlistment for seniors and graduates. Since such an important component of military enlistment is on-the-job training, it may be that for graduates—who have demonstrated a low propensity to attend college—expecting more education may make them more likely to enlist in the military than to pursue alternatives with fewer training opportunities. In contrast, seniors who indicate they expect more education would be less likely to enlist in the military and more likely to pursue educational alternatives.

Education costs and availability. A second type of factor expected to influence individual enlistment decisions is the cost and availability of higher education. Individuals who face higher education costs or lower availability will be more likely to enlist than choose an educational activity. Variables that Hosek and Peterson used to represent education costs and availability include *family income, number of siblings,* and *live at home.* Following Becker's theory of human capital (Becker, 1975), we expect that individuals with higher *family income* are more likely to be able to finance higher education. Holding all else constant, we expect individuals in families with a greater *number of siblings* to have less resources available to finance the education of each child. Finally, we expect individuals who *live at home* to be better able to pay for higher education since they are taking advantage of the economies of scale associated with living with their family and are not likely to be expected to contribute much in the form of rent or housing costs.

Civilian labor market opportunities. A third type of factor that Hosek and Peterson included in their model was measures of civilian labor market opportunities. They include variables reporting each individual's labor market opportunities, including *employment status, hourly wage, weekly hours of work, job tenure, time since last job,* and for graduates, *time since last attended school.* As they mention in their report, these variables are somewhat problematic as measures of civilian labor market opportunities because they represent a combination of labor market conditions and choices the individual has made such as how much to work and the quality of the job match. In addition, there is the problem that these variables are only observed for individuals participating in the labor force.

We expect the probability of enlistment to decline with an individual's wage rate. A higher *hourly wage* suggests that the value of working in the civilian sector will be relatively greater. Similarly, individuals with greater *weekly hours of work* and *job tenure* might be less likely to enlist given that they have demonstrated greater attachment to the labor force. However, for seniors, higher levels of these variables could indicate a lower likelihood of attending college, which could lean in favor of the respondent enlisting rather than attending college. The longer the *time since last job* and, for graduates, *time since last attended school,* the more likely is enlistment given that the individual has demonstrated a low value for the competing alternatives.

Race and ethnicity. The Hosek and Peterson model also includes variables that indicate whether a respondent is of Hispanic ethnicity or, if not Hispanic, then of black race. Blacks have been historically overrepresented among enlistees (see Binkin et al., 1982; Phillips et al., 1992). Some have reasoned that this is due to blacks having relatively fewer civilian educational and job opportunities than whites (Hosek and Peterson, 1985; Binkin and Eitelberg, 1986; and Phillips et al., 1992). In addition, the military is often perceived as a meritocracy that is not subject to the same degree of racial prejudice as the civilian world (see Segal (1989) and Segal, Bachman, and Dowdell (1978), for example). While no study has conclusively identified the reasons for black overrepresentation, several studies have found that even after controlling for other background variables that are likely to be correlated with race—such as mother's education, family income, number of siblings, and others—blacks are still more likely to enlist than observationally similar whites (Hosek and Peterson, 1985, 1990; Kilburn, 1994).

As discussed in Gorman and Thomas (1993), race is correlated with a number of other factors associated with enlistment, such as AFQT scores and family income. Therefore, to estimate the effects of race on enlistment net of these other characteristics, it is important to include these other factors as covariates in the model so that they are held constant when estimating the race effect.

Like blacks, Hispanics are not as successful in the labor market and do not attain as much education as whites (Bean and Tienda, 1987; U.S. Bureau of the Census, 1996). In contrast to blacks, however, Hispanics do not appear to be more likely to view the military as a superior alternative to the civilian sector. Studies of individual enlistment probabilities have found that holding other variables constant, Hispanics are less likely than whites to enlist (Hosek and Peterson, 1985, 1990; Kilburn, 1994).

Determinants of Military Behavior: Demand Factors

As discussed in Hosek and Peterson (1985), the demand side of the enlistment market includes a number of factors that cannot be studied with individual-level data. These include items that vary at the national level but not the local level, like national advertising, the effects of enlistment incentives like bonuses and educational benefits, Delayed Entry Program policy, recruiter management, and others. These factors are all taken as given.[1] Since the model used in this report explores variations across states but not across time, we examine only the relationship between individual enlistment decisions and demand factors that vary across states. The three such demand factors that we measure are *recruiter density,* the *market share of seniors and recent graduates,* and *AFQT category IV.* We now discuss each of these in turn.

Recruiter density. The recruiter density is the number of recruiters relative to the male youth population in an area. We expect the enlistment probability to rise if the youth is in an area with high recruiter density. In an area with higher recruiter density, the individual would be more likely to be contacted by a recruiter and possibly persuaded to enlist. We do not expect this relationship to vary substantially between seniors and recent graduates.

Market share of seniors and recent graduates. We also include a variable that measures the *market share of seniors and recent graduates.* Note that this is slightly different from the definition used by Hosek and Peterson, which was the proportion of current high school seniors and those who were seniors the previous June in an area's male youth population. Hosek and Peterson (1985) hypothesize that the greater the number of seniors in an area, the fewer the graduate enlistments. Because most seniors are contacted through recruiter visits, which do not vary much with the size of the senior population, a large share of seniors would lead to a large number of senior recruits in the total recruiting goal. As a result, the need to contact and recruit graduates would decline, and hence, the probability of a graduate enlisting would be lower the greater the share of senior and recent graduates.[2]

AFQT Category IV. By law (U.S.C. 10, Section 520), individuals scoring in the 10th through 30th percentiles on the AFQT—AFQT Category IV—are only allowed to enlist in limited numbers: no more than 20 percent of a fiscal year cohort can come from this category (see Kilburn et al. (1998) for further discussion of this issue). However,

---

[1]The net effect of these unmeasured factors is captured by the constant term in the probability equations.

[2]See Hosek and Peterson (1985, pp. 20–21) for a detailed discussion of this hypothesis, which is not repeated here.

DoD policy limits this number to no more than 4 percent.[3] In addition, individuals with scores in this range are less likely to meet the enlistment standards for many military occupational specialties (Eitelberg, 1988). As a result, we hypothesize that the enlistment probability of both seniors and graduates in this score range would be lower than for other individuals.

Applicants with children. No formal prohibitions preclude enlisting recruits who have children. However, the services typically do not enlist single parents who have custody of minor children or married individuals who have a large number of children. This leads us to hypothesize that individuals with children would have a lower probability of enlisting than would their contemporaries with no children, particularly if they were single.

---

[3]Correspondence from Dr. W. S. Sellman, Director, Accession Policy, OASD (FMP) (MPP).

Chapter Three

CHANGES IN THE RECRUITING ENVIRONMENT

Given the model of individual choice outlined above, what reason do we have to reestimate it with more recent data? That is, what has changed since 1980 that leads us to question the applicability of the Hosek and Peterson estimates to the current environment? A variety of factors have changed in the interim—factors that we expect would have affected both the individual characteristics and choice attributes, or X's in the model, and factors that would have affected the coefficients, or β's in the model. We briefly summarize these factors here, indicating likely ways they would influence the estimates.

We maintain the theoretical approach of Hosek and Peterson outlined in the previous chapter. In that model, recruit supply and demand factors interact to yield recruiting outcomes. The effects outlined here are *ceteris paribus*—that is, they describe a change elicited by one variable at a time while holding constant all other factors in the model. Note that like Hosek and Peterson, we only have one year of data available and so are not able to explore across time variation in variables. Hence, we do not explicitly test whether the changes discussed in this chapter have specific effects on the model in our later empirical chapter. The discussion here is meant to be a survey of some of the possible factors that changed between the early 1980s and collection of the NELS data that might have influenced the relationship between individual characteristics and enlistment probabilities.

In general terms, a change that is related to one of the variables we include in the model is likely to result in a change in the levels of the variable or in the estimate of the coefficient for that variable. For changes related to factors that are not explicitly incorporated via one of the explanatory variables in the model, the effect is likely to be a change in the intercept term. Again, we group these factors according to whether they influence recruit supply or recruit demand. First, we discuss the factors associated with recruit supply.

SUPPLY FACTORS

Shrinking Youth Cohort

One of the factors included in every aggregate model of recruiting is the size of the youth cohort. One estimate is that growth in the youth cohort of 10 percent is associated with a 2.4 percentage point increase in the number of high-quality recruits, all else held constant (Asch and Orvis, 1994). In 1980, the number of 18-year-old males in the United States was near a peak for this century (see Klerman and Karoly, 1994). By 1992, the number of 18-year-old males had shrunk to approximately three-quarters the size of the 1980 cohort (U.S. Bureau of the Census, *Current Population Reports,* various issues). The decline in the female youth population was similar. Holding other factors constant, including demand factors, a smaller youth cohort would make an individual in that cohort more likely to enlist. In estimates from only one cohort, this would imply that given the same characteristics as a youth from a larger cohort, the youth from the smaller cohort would be more likely to enlist. Given that this is not a factor explicitly included in our model, this change is likely to result in a larger estimate of the intercept.

Greater Share of Minorities and Immigrants in Youth Cohorts

At the same time that the youth population was shrinking, its composition was also changing. Since the inception of the All-Volunteer Force (AVF), blacks have been overrepresented relative to their share of the population while other minorities, most notably Asians and Hispanics, have been underrepresented relative to their share of the population (Office of the Assistant Secretary of Defense, 1996). Hosek and Peterson (1985, 1990) report that for both seniors and graduates, black youths were more likely to enlist than white youths. They found that for Hispanics, their estimate for seniors was marginally significant and indicated that Hispanics were more likely than whites to enlist and, for graduates, that there was no statistical difference between the enlistment probabilities for whites and Hispanics.

Some recent evidence suggests that these trends might be changing. The representation of blacks in the armed forces has declined in recent years. While more than 20 percent of accessions were black during the late 1980s, the fraction of new recruits who were black has not exceeded 18 percent since 1990 (Office of the Assistant Secretary of Defense, 1996). At the same time, the representation of Hispanics has increased. Prior to 1989, Hispanics had never comprised more than 6 percent of accessions. In 1990 they made up 7 percent of accessions, and by 1995 this had risen to 9 percent (Office of the Assistant Secretary of Defense, 1996). In addition, other racial and ethnic groups have begun to make up a larger share of the enlisted accession population (Office of the Assistant Secretary of Defense, 1996). Other evidence that calls for a reexamination of minority enlistment patterns comes from analysis of the Youth Attitude Tracking Survey (YATS). Orvis et al. (1996) report that while there has been a recent downward trend in enlistment propensity among all youth, this decline has been particularly acute among black youths. Both evidence on the representation of blacks among accessions and the falling enlistment propensity among blacks suggest that we would estimate lower coefficients on the black indicator variable in our probability equations. We anticipate little change in the coefficient on the Hispanic indicator variable.

In addition to the fact that the racial and ethnic composition of the workforce has changed, there have also been changes in the percentage of immigrants in the workforce. In 1980, 6.6 percent of the male workforce was immigrants. By 1992, this percentage had grown to just under 10 percent (Schoeni et al., 1996). Furthermore, the growth in the representation of immigrants was greatest among workers with less than a college education. The largest increases in immigration were among groups of Hispanic and Asian descent (Schoeni et al., 1996) and these groups tend to be underrepresented in the military (Office of the Assistant Secretary of Defense, 1996). The increase in the representation of immigrants in the workforce is likely to contribute to the rise in the fraction Hispanic in 1992 relative to 1980. Since we know little about the enlistment behavior of immigrants or their descendants, it is unclear what effects the rise in immigration would have on coefficients in the model.

Increase in College Attendance

The rate of school attendance among 18- and 19-year-olds grew dramatically over the period 1980 to 1992. In 1980, 46 percent of 18- and 19-year-olds were enrolled in school. By 1992, over 61 percent of individuals in this age range were enrolled—nearly a one-third increase (U.S. Bureau of the Census, 1994). Note, however, that this increase was not equally distributed across racial and ethnic groups. The rate of college attendance among whites during this period largely followed the overall trend, while the growth in Hispanics' college attendance exceeded the overall trend and blacks' college attendance did not grow as much as that of the other groups. A corollary to the rise in college attendance is the fact that the returns to skill in the labor market have increased over the period 1980 to 1992. It is well documented that during this period, the wages of more-educated individuals have risen relative to individuals with a high school degree or less (see the review in Levy and Murnane, 1992). We discuss this issue again below in the context of military wages.

It is likely that individuals who can attend college would be eligible for enlistment. Therefore, this rise in college attendance may reduce the likelihood that eligible youths enlist in the military. This change would show up in our estimates as a drop in the intercept estimate. This change is also likely to result in a smaller high school graduate pool from which the military can recruit. Recall that Hosek and Peterson include in the graduate sample individuals who have graduated from high school but who are not enrolled in school.

Another possibility that would be consistent with this change in college attendance would be that the coefficients in the model did not change, but rather that the characteristics of the population or the attributes of the choices changed. For example, say that in the past we found that individuals whose mothers had more education were less likely to enlist and more likely to attend college than individuals whose mothers had less education. If mother's education grew over the period, we might observe that individuals were less likely to enlist and more likely to attend college while none of the coefficient estimates changed. In other words, the observed increase in college attendance would be consistent with both (1) a change in the intercept with no change in the effect of mother's education or in the levels of mothers' education, and (2) no change in the intercept, no change in the effect of mother's education, and growth in mother's education.

Finally, given the rise in the returns to education, it may be the case that there has been a change in the relationship between enlistment probability and some of the variables in the model that represent educational alternatives. For instance, family income is included in the model because it is believed that families with higher income are better able to finance children's education and therefore their children would be less likely to enlist. If the returns to education have grown substantially, it may be the case that individuals are more willing to take out loans or consider alternatives to family financing of education. If this were the case, we may observe a decline in the size of the coefficient on family income in our model.

Youth Aptitudes Have Risen

In another report that is part of this project, Kilburn et al. (1998), we estimate that slightly more high school seniors are "high quality"—AFQT CAT I–IIIA—in 1992 than in 1980. In addition, we found that slightly fewer seniors were ineligible due to low AFQT scores in 1992 than in 1980. The changes in youth aptitudes over the period varied substantially by gender and race/ethnicity, however. Whites experienced little growth in scores; the increase in blacks' scores was dramatic, and the rise in Hispanics' scores was sizable but less than that of blacks. We estimated that while 8 percent of black seniors scored CAT I–IIIA in 1980, almost 20 percent did so in 1992. For Hispanics, 21 percent scored CAT I–IIIA in 1980; by 1992, 27 percent scored in this range. We also found that a large part of the gains in black and Hispanic test scores was due to growth in females' test scores rather than males' scores.

These results suggest that a higher fraction of minority youths would be eligible for enlistment in 1992 than in 1980. This implies that recruiters are able to draw from a larger pool of potential applicants. This change is likely to manifest itself in two ways. First, we would expect to see a difference in the characteristics of the 1992 and 1980 cohorts. The 1992 cohort should have fewer individuals in the *AFQT*

Category IV, and should have higher fractions of blacks and Hispanics in the eligible pool. Both of these patterns hold true when comparing the NELS data to the Hosek and Peterson sample. Second, holding all else constant, this translates into a larger recruit supply, which should result in a smaller intercept estimate.

Increase in Military Pay and the Decline in Real Wages of High School Graduates

Another factor that is different for individuals considering enlistment in 1980 and those considering enlistment in 1992 is the level of military pay relative to civilian pay. Over the period FY80–82, enlisted personnel received a 33 percent pay raise to overcome a perceived gap between military pay and civilian wages (see Office of the Secretary of Defense, 1991, pp. 30–31). At the same time that real military pay for enlistees was rising (Hosek, Peterson, and Heilbrunn, 1994), the civilian pay of similarly educated civilian workers—those with a high school degree but no college—began to slide in real terms. In real terms, the average wages for men 25 to 29 years old with a high school degree fell by approximately one-tenth from 1980 to 1992 (Klerman and Karoly, 1994). Moreover, the gap between workers with a college degree and those with less education has been widening since the early 1970s.

These changes in wage structure all work together to make the military a more attractive employment option in 1992 than in 1980 for individuals who are unlikely to pursue education beyond a high school degree. This could have the effect of raising the probability of enlistment for individuals toward the lower end of the test score distribution, since they are less likely to attend college. This might imply a smaller coefficient on the AFQT percentile variable but a larger one on the AFQT Category IV variable, holding all else constant.

Desert Storm and Rise in Deployments

Operation Desert Storm took place during the senior year of members of the NELS sample. In addition to this highly publicized deployment were a number of smaller deployments over the period 1992–1994 and a growing operating tempo overseas (Hosek and Totten, 1999). In contrast, individuals in the NLSY sample were considering enlistment in 1980, a period of relatively modest deployments following the cessation of the Vietnam War. There has been speculation that higher operating tempo and more deployments have contributed to a perception of greater risk of military life among youths (Office of the Assistant Secretary of Defense, 1995), and there is evidence of a downturn in propensity immediately after the Gulf War (Orvis et al., 1996). On net, the perception of a rise in deployments could either lower or raise individuals' probability of enlisting and, in our model, lower or raise the intercept estimate. On one hand, more deployments could be associated with greater risk or disruption of service members' personal lives, which might lower enlistment rates. On the other hand, deployments might be desirable to potential recruits in that they increase opportunities to carry out the services' mission and to serve the country. Another possibility is that the effect of deployments on enlistment would depend on youths' perceptions of the type of deployments. This is similar in spirit to findings that service members with short deployments are more likely to reenlist than those with no deployments, but that as the length of deployments grows, individuals are less likely to reenlist (Hosek and Totten, 1999).

DEMAND FACTORS

In addition to the changes in factors associated with the supply of recruits, there have been changes in a number of demand-related factors.

Smaller Recruiting Requirements

With the drawdown in the size of the military that took place over the period 1990–1993 came a more than proportionate cutback in recruiting requirements. These reductions were dramatic: in the Army, about 130,000 recruits were inducted in 1979 but only about 75,000 entered in 1992. Across all the services, non-prior-service accessions fell from over 350,000 in 1980 to just over 200,000 in 1992 and 1993. This implies that all else equal, a youth in 1992 would be less likely to enlist than a youth in 1979. Such an effect would lead to a lower coefficient estimate for the intercept term.

Rising Recruit Quality

At the same time that total non-prior-service accession requirements were shrinking, the quality of recruits was rising as measured by AFQT scores. In 1992, both the fraction of high-quality recruits and the absolute number of high-quality recruits were substantially larger than in 1980. In 1980, about 49 percent of new recruits scored in the upper half of the AFQT distribution. By 1992, 75 percent of new recruits scored in the upper half (Office of the Assistant Secretary of Defense, 1996). This would favor the chances of high-quality youths enlisting in 1992 relative to their counterparts in 1980. Note that accompanying this would be a concomitant reduction in the chances of observing a low-quality youth enlisting in 1992 relative to 1980. This would result in a larger estimate for the coefficient on AFQT score because a higher AFQT score would raise the probability that an individual enlists in 1992 even more than it did in 1980.[1] Individuals with extremely low AFQT scores should be even less likely to enlist in 1992 than individuals with low scores in 1980. Therefore, we expect the new coefficient estimate for the variable *AFQT Category IV* to be lower than that estimated by Hosek and Peterson.

Expansion of Women's Roles in Military

The expansion of women's roles in the military is likely to have raised the relative odds of women enlisting and lowered the relative odds of men enlisting. Two types of policies are explicitly directed toward raising women's participation. One is that more military occupational specialties (MOSs) have become open to women over the last decade and a half (Harrell and Miller, 1997). With more opportunities to serve, women's enlistment is becoming less demand-constrained. The second policy aimed at boosting women's participation has been recruiting goals for female recruits (see Oken and Asch, 1997). Combined, these changes are likely to raise women's likelihood of enlisting but also to reduce men's likelihood of enlisting. Therefore, our estimates of the intercepts in men's equations ought to be lower as a result.

---

[1]Note that below we suggest an alternative hypothesis for the relationship between AFQT scores and enlistment probability that is related to a different change in the environment.

Cutbacks in Advertising

The services cut back on advertising expenditures during the drawdown period (Asch and Orvis, 1994). It has been shown that advertising has both an immediate and lagged effect on the number of high-quality enlistments (Dertouzos and Polich, 1989). Given that advertising campaigns may take a long time to have an impact on enlistment decisions, it is not clear what short-run downturns in advertising dollars imply for short-run enlistment numbers. Hence, the impact of the overall reduction in advertising expenditures during the drawdown period on the number of enlistments is unclear.

Reduction in Number of Recruiters

As part of the drawdown, when recruiting requirements dipped, the services reduced the number of recruiters (Asch and Orvis, 1994). Aggregate models of recruiting find that the number of recruiters is an important determinant of the high-quality recruit yield. For instance, a 10 percent increase in the number of recruiters could result in close to 6 percent more high-quality recruits (Asch and Orvis, 1994). The number of production recruiters in the Army fell approximately 5 percent between 1980 and 1992, from 4,708 to 4,463. The effect of this trend on individual enlistment probabilities is likely to be a decline in the intercept of the 1992 estimates relative to the 1980 estimates. Note that an overall drop in the number of recruiters may be related to the estimate of the coefficient on the *recruiter density* variable. This coefficient measures the difference recruiter density makes across geographic units at a point in time. If the average number of recruiters nationwide dropped, it may mean that the effect of adding an additional recruiter in a geographic area has risen due to diminishing marginal returns. That is, the lower the number of recruiters, the greater the marginal benefit may be of adding an additional recruiter. Hence, we may also observe an increase in the coefficient on *recruiter density* as a result of the decline in the number of recruiters between 1980 and 1992.[2]

---

[2]For a review of the influence of recruiter density on enlistment in aggregate models, see Murray and McDonald (1998).

Changes in Recruiter Management

The four services use a variety of incentive plans and quota systems to encourage recruiters to be most productive. For example, incentive plans such as the Navy's Freeman Plan could alter the relative odds of recruiting a high-quality versus a low-quality recruit (Asch, 1990). Other recruiting policies may affect the likelihood that females or individuals from particular ethnic groups enlist (see Oken and Asch, 1997). In addition, there are other players in the recruiting system such as job counselors, who are also subject to incentive plans (Asch and Karoly, 1993). Recruiter management has a complicated history: it took a different course in each of the four services, and the nature and structure of recruiter incentives changed many times between 1979 and 1992 (see Oken and Asch, 1997). Due to the diverse nature of the changes in recruiter management over the period 1980–1992, we do not have specific hypotheses on how these changes would have influenced the estimates of the individual enlistment model. While we do not explicitly measure recruiter management variables nor include them in our model of individual enlistment decisions, it is worth noting that this is an area warranting additional investigation.

Altering Enlistment Incentive Programs

A final demand-side factor that has changed over the period between the Hosek and Peterson studies and our study is educational assistance programs and enlistment bonuses offered to potential recruits. The basic idea behind the educational assistance programs is that offering to pay for postservice education will be especially effective in attracting high-quality recruits (Fernandez, 1982). Educational benefits were introduced and tested in the early 1980s in response to difficulties in attracting the desired levels of high-quality recruits (Asch and Dertouzos, 1994). In the mid-1980s, Congress created the Montgomery G.I. Bill and the Army College Fund. Enlistment bonuses are often targeted to high-quality youth willing to sign up for hard-to-fill occupations, or they may be used to lengthen term-of-service agreements or accomplish other specific recruiting objectives. The enlistment bonus program was significantly expanded in 1982 (see Asch and Orvis, 1994). In general, it appears that more enlistment incentive programs were available to the 1992 cohort than to the 1980 cohort, in terms of both enlistment bonuses and educational incentives. Given that most of these incentives were directed at higher-quality individuals, we hypothesize that these programs would have the effect of further raising the coefficient on the AFQT variable.

Higher Crime Rate Among Youth

One of the eligibility requirements for enlistment is passing a moral screen. This includes having been law abiding. Potential recruits may have had increasing difficulty meeting this enlistment standard between 1980 and 1992 due to an increase in the arrest rate among juveniles for violent crimes and weapons violations. The rate of arrest for these two types of violations rose by one-third between 1980 and 1992, from 5.8 per one thousand to 7.7 per one thousand (U.S. Bureau of the Census, 1994).[3] Counteracting this trend in youth arrests for violent crimes are the trends for youth arrests on drug charges, which actually declined over the period 1980 to 1992. In 1980 the arrest rate for drug violations was about 9.5 per thousand while the rate was about 5.5 per thousand in 1992 (U.S. Bureau of the Census, 1994). On net, since recruits are more likely to be able to get a moral waiver for drug violations than for violent crime offenses, we expect the rise in violent crime arrests to dominate the drop in drug arrests. Since there are no variables corresponding to arrests in the Hosek and Peterson model, we expect this change would manifest itself in a small, perhaps imperceptible, decline in the intercept.

INNOVATIONS IN THE SPECIFICATION

In addition to estimating an individual enlistment model that contains the same variables that Hosek and Peterson used, we also estimate a model that adds some variables and omits some variables. Some of the innovations we make are suggested by the changes enumerated above that have taken place since Hosek and Peterson estimated their model. Other innovations in the specification are suggested by economic theory. First we describe the changes to the

---

[3]A juvenile here is defined as a person between the ages of 10 and 17.

specification that are suggested by changes that took place between 1980 and 1992.

The first innovation we make in the specification is to add an in-state college tuition variable. We include this variable because the rise in college attendance rates suggests that characterizing the college opportunities of potential recruits is an increasingly important dimension of the enlistment decision. The probability that an individual enlists should be positively related to the college costs that individual faces. Different states subsidize public higher education to varying degrees, so the tuition at these institutions differs. Prior studies have found that military enlistment is higher in states with lower college tuition (Kilburn, 1994) and that college attendance is higher in states with lower tuition (Kane, 1994). In 1992, the year most of the NELS respondents would have entered college, average in-state tuition ranged from $830 to $5,300 per year. Given that college attendance rates rose between 1979 and 1992, it may be that the cost of education is even more important for the NELS respondents than for the NLSY sample Hosek and Peterson used. While we cannot evaluate whether college costs became more influential for enlistment rates during this period, we examine whether this variable influences enlistment probabilities for the NELS sample.

We also add a variable indicating whether a respondent is from a family likely to have recently immigrated to the United States. As discussed earlier, immigrants are one of the fastest-growing segments of the youth population, and little is known about their enlistment behavior. We include a variable that specifies whether English is the respondent's second language to indicate whether he or she is likely to have come from an immigrant family. While it is not clear what sign the coefficient on this variable would have, one piece of evidence is at least suggestive. Many immigrants in the 1980s and 1990s were Hispanic, so the immigrant groups may exhibit enlistment behavior that is similar to that of Hispanics.

As discussed above, another change that has taken place since the late 1970s is the growth in youth arrests for violent crimes along with a declining arrest rate for drug violations. Given that the military has moral requirements that limit the enlistment opportunities of individuals with arrest records and that the military prohibits drug use, the trend in violent crime arrests could be reducing recruit supply while the trend in drug arrests could be increasing potential recruit supply. Two variables from the NELS are included in our new specification to proxy for behaviors that violate the military's moral code. The first variable reports whether the individual or a close friend has ever been arrested, and the second indicates whether the respondent has ever used marijuana. In addition to these two variables, we include variables that indicate when they are missing. We hypothesize that the coefficient estimate for both of these variables should be negative, indicating that individuals in those categories would be less likely to enlist.[4]

The last variable we add to the new specification is one that indicates whether either of the respondent's parents was in the military at the time of the 1992 survey. Economic research on occupational choice has found that children of individuals in a particular occupation are more likely to choose that occupation than are children whose parents are not in that occupation (Reville, 1996). Research on military occupational choice reports that military careers are even more likely to follow family lines than civilian occupations (Thomas, 1984). Therefore, we expect that children whose parents were in the military ought to be more likely to choose to enlist, so we should find a positive coefficient on this variable.[5]

Next we make a few innovations to the specification that are driven by theoretical considerations. First, we include AFQT category indicators rather than AFQT percentile score. We make this change because tabulations of enlistment across the AFQT distribution show that enlistment does not uniformly rise or fall as AFQT percentile increases. Instead, for seniors, individuals toward the middle of the distribution are more likely to enlist, implying that the probability of enlisting first rises as AFQT percentile goes up from zero and then falls as the AFQT percentile reaches the highest categories (see Table

---

[4]Note that it is unclear how reliable self-reporting of criminal behavior and drug use is. The missing indicators for these variables may indicate unwillingness to report the behavior.

[5]Note that our variable does not indicate whether the respondent's parents were ever in the military, only whether one of the parents was in the military at the time of the 1992 survey.

3.1).[6] Including the AFQT categories rather than AFQT percentile allows us to capture this nonlinear relationship between AFQT and enlistment. For graduates, the enlistment rate is highest for CAT I and CAT II, and lowest for CAT IV and CAT V.

The last innovation we make to the Hosek and Peterson specification is to remove most of the variables related to current working behavior. We do this because for the NELS sample, some of the individuals who enlisted did not report their prior working behavior, which may bias our results for these variables.

ESTIMATING A THREE-CHOICE MODEL OF ENLISTMENT

Last we estimate a model of enlistment choice that includes two other activity alternatives—college or work/other—rather than just one alternative choice—not enlisting.[7] There are two advantages to

Table 3.1

Percent of Seniors and Graduates Enlisting as of 1994, by Estimated AFQT Category

	Seniors		Graduates	
Estimated AFQT CAT	Number in AFQT CAT	Percent Enlisting[a]	Number in AFQT CAT	Percent Enlisting[a]
I	343	0.86	22	30.05
II	1,725	3.38	306	15.02
IIIA	954	4.94	273	7.70
IIIB	1,406	6.79	600	8.93
IV	1,550	4.27	1,044	4.43
V	332	4.54	244	1.04

[a]Weighted NELS data.

[6]Our AFQT measure is an estimate of the individual's AFQT score in the spring of senior year rather than the true AFQT score the person would have obtained at the time he was considering enlistment. This is why we have enlistments coming from low AFQT categories even though the number of enlistments coming from these categories is constrained by legislation.

[7]Note that there are other choice sets we could have specified in addition to the three-choice model. For example, we could have allowed four choices, adding reserve enlistment to the three choices described above.

using a trivariate instead of bivariate model of enlistment. First, the theory discussions above clearly indicate that some explanatory variables are expected to influence the likelihood of enlisting because they are associated with greater benefits or cost of going to college, while others are expected to influence enlistment because of their association with greater benefits or costs of working. Examples of covariates that would be expected to be related to college-going as an alternative include the family's income and the number of siblings. Examples of variables that are associated with the working alternative are the unemployment rate and the female labor-force participation rate. Using a model with college and working/other as separate alternatives more clearly indicates whether the variables have the observed effect on enlistment via the theoretical relationships posited above.

Another advantage of the trivariate model is that we expect some explanatory variables to raise the likelihood of enlisting relative to college but lower the likelihood of enlisting relative to working/other. If this is the case, in the bivariate model the estimated coefficient for this variable is likely to tend toward zero, since the alternate choice includes both college and work/other. Using the trivariate model will enable us to evaluate whether the variable has the expected effect relative to the two enlistment alternatives. An example of such a variable is the respondent's AFQT score. While we would expect the average AFQT scores of enlistees to be lower than the average AFQT scores of college attendees, we would expect the average scores of enlistees to be higher than those of youths who work right after high school. Hence, while the coefficient estimate on AFQT score might be zero in the bivariate model—because the negative effect of the college alternative and the positive effect of the work alternative cancel each other out—it may be significant and of opposite signs in the trivariate estimate of college and work as alternatives to enlistment.

Chapter Four

NELS DATA

The National Education Longitudinal Study (NELS) follows a representative sample of individuals who were eighth graders in 1988, obtaining information on high school, postsecondary education, work, family formation, and background characteristics. The 1988 sample was selected using a two-stage probability strategy. In the first stage, approximately 1,000 public and private schools were selected from the universe of about 40,000 schools containing eighth graders. In the second stage, random samples of 24–26 students per school were selected. Also included in the sample are a parent, the school principal, and two teachers for each selected student. The study oversamples Hispanic and Asian students.

The NELS interviewed respondents in the Base Year (1988), a First Follow-Up (1990), a Second Follow-Up (1992), and a Third Follow-Up (1994). In most of the follow-ups the school samples were "freshened," a process that adds students to compensate those who have dropped out, left to study abroad, or emigrated, so that the sample remains representative of a random sample of students in a particular grade level. Hence, despite the fact that some students from each earlier wave of the study were no longer in school, the First Follow-Up is representative of students enrolled in 10th grade in the spring of 1990, and the Second Follow-Up is representative of students enrolled in 12th grade in the spring of 1992. The Third Follow-Up was not freshened.

Each interview includes a student questionnaire for individuals still in school, a dropout questionnaire for respondents no longer in school, a teacher questionnaire that asks teachers about specific respondents as well as class and school climate information, and a school questionnaire to obtain characteristics of the school. The student questionnaire collects information on family background, school activities, plans for the future, and other characteristics. The Second Follow-Up also reports the respondent's score on cognitive tests in the areas of reading, math, science, and social science. These tests are unique to the NELS and were designed to measure the acquisition of aptitudes appropriate for the 12th grade. This follow-up also asks seniors if they have enlisted in the military.

The third Follow-Up surveys respondents two years after high school graduation. This questionnaire asks respondents to report on education, work, family formation, and other activities over this two-year period. We can identify which respondents enlisted during the period using both contemporaneous and retrospective questions. Hosek and Peterson (1985, 1990) distinguished between seniors and graduates. Using the questions in both the Second and Third Follow-Ups, we can also distinguish between individuals who enlisted while seniors and those who enlisted after graduating.[1]

Because it contains a large representative sample of youth who are seniors in high school and followed after graduation, this data set is well suited to conducting the present study. Other strengths of the data set are that it follows respondents after they enlist, it contains a test score that can be used to mimic the AFQT, and it contains a rich set of background variables like those used in the Hosek and Peterson model.

Despite the advantages of these data, there are numerous flaws as well. First, no oversample of enlistees is available to augment the data; the Hosek and Peterson studies were able to use the AFEES[2] data in conjunction with the random NLSY sample. As a result, there are relatively few enlistees in the data, somewhat reducing the power of our statistical tests. In particular, even if the true effects were the

[1]While the NELS sample will allow us to study the enlistment behavior of one cohort of high school seniors, it does not provide a comprehensive view of enlistment behavior: individuals as old as 35 years are eligible to enlist, as are those who never reached the 12th grade. However, most individuals who enlist do so at the ages at which we observe the NELS sample: nearly three-quarters of nonprior accessions are age 20 or younger (Office of the Assistant Secretary of Defense, 1995).

[2]Armed Forces Entrance Examination Stations.

same as in the Hosek and Peterson studies, because our samples are smaller, we would be less likely to find statistically significant effects. Second, the variables in the NELS are not always defined in exactly the same way as the variables in Hosek and Peterson were, making our comparison of results less precise than we would prefer. For example, the NLSY asked for family income using a continuous variable, whereas the NELS asked for family income in categories of income levels. Nevertheless, for all the variables in the Hosek and Peterson specification, we are able to find exact or close matches in the NELS.

Among individuals in the NELS who were seniors in 1992, we count someone as having enlisted if they answer yes to the question "Have you enlisted in the military?" We find that 302 of our 7,671 senior males enlisted, or 4.28 percent (weighted) of our senior sample. This is slightly higher than the enlistment rate reported for 18-year-olds in 1992 by DoD (Office of the Assistant Secretary of Defense, 1995). Note that our sample only includes individuals who are still in school in February of their senior year and does not include dropouts. Since dropouts are ineligible for enlistment or at least are allowed to enlist only in limited numbers, it is not surprising that we find an enlistment rate for seniors that is higher than the enlistment rate for 18-year-olds including both seniors and dropouts.

As in the sample Hosek and Peterson used, the graduate sample includes only high school graduates who are not enrolled in school. Individuals can be currently serving in the military or engaging in some other nonschool activity. We count someone as being enlisted if the person meets one of the following conditions in the 1994 wave of the questionnaire:

- Answers yes to the question "Are you now serving on active duty in the Armed Forces?"
- One of the occupation codes for jobs held since graduating from high school is "Military" and the branch of the military was either Army, Navy, Marines, or Air Force.

We find that 255 out of 3,562 individuals in our graduate sample enlisted in the military, or a weighted 7.17 percent. We do not have a comparison group that is similarly defined, making it somewhat difficult to assess the reasonableness of this estimate. However, in

1994, the year our sample was reinterviewed, the DoD reports an enlistment rate of 2.1 percent for 19-year-olds and 1.3 percent for 20-year-olds (Office of the Assistant Secretary of Defense, 1995). Given that roughly half of individuals 19 and 20 years old were enrolled in school and hence not eligible to be in our graduates sample, this would imply that our enlistment rate should be close to 6.8 percent,[3] which is close to the enlistment rate of graduates in our data.

While the NELS data enable us to compare more recent estimates of individual enlistment models to those obtained by Hosek and Peterson, these comparisons have some limitations. First, note that our graduate sample is slightly different from the one used by Hosek and Peterson in that they included individuals in a range of years past high school graduation and who were seniors in different years. Our sample only includes individuals who were seniors in 1992 and interviews them all at a point about two years after graduation.

Second, our local labor-market variables are based on state means rather than the county means used by Hosek and Peterson. We are likely to obtain weaker relationships between these variables and individual enlistment decisions than those estimated for Hosek and Peterson's more refined labor-market measures.

Second, when comparing the NELS results with the Hosek and Peterson (1990) results, it is important to remember that enlistment is much less common in the NELS than in the choice-based sample of Hosek and Peterson. While less than 5 percent of the NELS senior sample and less than 7 percent of the NELS graduate sample enlisted, over 87 percent of seniors in the Hosek and Peterson sample enlisted, as did about 78 percent of their graduate sample. This difference in enlistment rates across the two samples has implications for the relative precision of estimates using the two samples. The result is that our regression coefficient estimates have larger standard errors, smaller t-statistics, and are less likely to be statistically significant than the Hosek and Peterson estimates. Therefore, even more so than in most regression analyses, when comparing the NELS results and the Hosek and Peterson results, it is important to compare the results of the two studies using statistical tests that incorporate

---

[3]6.8 percent = (2.1 percent + 1.3 percent)/one-half.

standard error differences rather than by noting differences in statistical significance. To compare results of the two studies below, we present statistical tests of differences. Technical details of the tests are provided in Appendix B.

We do not include women in our study because there are too few female enlistees to obtain statistically significant results. Less than 1 percent of women in any cohort enlist, and given our sample size from the NELS, we had only 36 female enlistees in our senior sample and 54 female enlistees in our graduate sample.

Chapter Five

RESULTS

This chapter reports our estimates of the models discussed above. First, we present estimates of a bivariate model of enlistment that replicates the specification in Hosek and Peterson (1990). We compare our coefficient estimates to theirs and compare the characteristics of our sample and their sample. Then we estimate the amount of differences in enlistment outcomes that can be attributed to characteristic versus coefficient changes over the period 1980 to 1992. Second, we estimate a bivariate model of enlistment that includes additional covariates and deletes some covariates from the Hosek and Peterson (1990) specification. Finally, we estimate a multivariate model of the choice to enlist versus attend college or work in the labor force.

REPLICATING HOSEK AND PETERSON

We first estimate a specification that replicates that of Hosek and Peterson (1990) as closely as our data will allow. Due to the way questions were asked in the NELS, some of the variables we use are slightly different from those Hosek and Peterson used. For example, we generate a continuous mother's education variable with the categorical mother's education variable from the NELS because Hosek and Peterson used a continuous mother's education variable from the NLSY. Appendix Table A.1 includes the complete set of covariates along with their definitions and how they might differ from those used by Hosek and Peterson. Table A.2 reports the means and standard deviations of these variables.

Our coefficient estimates for the senior and graduate models using the Hosek and Peterson specification are presented in Table 5.1. The first two columns report the results for seniors, and the second two columns report the results for graduates. Along with the coefficient estimates and their standard errors, we report the log-likelihood, a Chi-square statistic for a test between our model and a model with just a constant, and the "pseudo R-square," which is defined as one minus the ratio of the log-likelihood of our model over the log-likelihood of the constant-only model. We also report a scale factor, a scalar multiple that converts each coefficient estimate into the marginal effect of a change in a variable on the predicted probability.[1]

In the senior model, the significant variables tend to be those that theory ties most closely to college attendance alternatives, such as AFQT score, mother's years of schooling, family income, and number of siblings. In the graduate model, the significant variables tend to be those most associated with work alternatives, such as weekly hours employed, employment status, and duration of unemployment. In both the senior model and the graduate model, many fewer variables are significant than was the case in Hosek and Peterson (1985, 1990). As mentioned above, this is likely to be due in part to the fact that the NELS data have many fewer enlistees than the Hosek and Peterson sample.

Although the model is able to identify some individual characteristics that significantly raise the likelihood of enlisting, the predictive power of the model is relatively low. The high Chi-square value reported in the table indicates that the model has significantly better fit

[1]For the logit, the marginal effect of a change in a variable X on the predicted probability is

$$\frac{\partial F(\hat{\beta}'X)}{\partial X} = \left\{ \frac{\partial F(\hat{\beta}'X)}{d(\hat{\beta}'X)} \right\} \hat{\beta} .$$

The scale factor is this term in brackets and hence produces the marginal effect for any variable by multiplying the estimated coefficient for that variable by the constant scale factor (see Greene, 1990). The value of the scale factor depends on the point of evaluation; we use the mean value of X.

Table 5.1

Logistic Regression Estimates of Enlistment Probability Using Hosek and Peterson Specification, for Seniors and Graduates

	Seniors		Graduates	
Variable	Coefficient Estimate	Standard Error	Coefficient Estimate	Standard Error
Black	–0.4597	0.3113	0.2200	0.2854
Hispanic	–0.6047*	0.3374	0.2396	0.3633
Age 16 when senior	—	—	–0.9093	1.2091
Age 17 when senior	0.0726	0.1828	0.0493	0.2220
Age 19+ when senior	–0.3507	0.2923	–0.4089	0.4035
AFQT score (31–99)	–0.0134*	0.0080	0.0117	0.0095
Category IV indicator	–1.0251**	0.4415	0.2919	0.5801
AFQT score missing	–1.0022**	0.4582	0.7478	0.5905
GED	—	—	–0.1084	0.4645
Mother's years of schooling	–0.1351***	0.0387	0.0280	0.0736
Mother worked	0.2744	0.2167	0.1877	0.3995
Family income (in $ thousands)	–0.0186**	0.0088	–0.0154*	0.0090
Family income < $5,200	–0.2190	0.3838	–1.5747***	0.3816
Family income missing	–1.0522**	0.4881	–1.0555***	0.3826
Number of siblings	0.1268**	0.0560	0.0938	0.0711
Missing number of siblings	0.3306	0.2926	0.0266	0.4077
Lives at home	–0.1202	0.2764	–1.7779***	0.2836
Missing lives at home	0.7312	0.5436	—	—
Hourly wage (natural log)	–0.6430	0.8161	–0.0469	0.1505
Wage < $2.25/hour	–1.2128*	0.7363	—	—
Hourly wage missing	–1.1772	1.4171	4.2439***	0.7710
Weekly hours currently employed	0.0135	0.0144	0.0722***	0.0109
Months employed (natural log)	–0.1165	0.0879	–0.0722	0.1630
Not currently employed	–0.1376	1.6231	3.1304***	1.2218
Weekly hours not currently employed	0.0127	0.0078	0.0632***	0.0113
Missing hours not currently employed	—	—	4.2066***	0.7177
Months not employed	–0.1051*	0.0620	–0.3601***	0.1013
Not employed in last 12 months	–0.3204	0.9653	4.9063***	1.9818
Unemployment rate in county	0.1606*	0.0863	0.0860	0.1146
Share of seniors and recent graduates	2.3461	1.7828	–1.6950	2.9097
Percent of population that is black	–0.0690	1.2433	1.2234	1.3002
Percent of population that is Hispanic	5.3661	7.0203	–6.6640	9.5073

Table 5.1 (continued)

	Seniors		Graduates	
Variable	Coefficient Estimate	Standard Error	Coefficient Estimate	Standard Error
Percent of labor force that is female	–0.0213	0.1070	0.0060	0.1379
Per-capita personal income	–0.0197	0.0123	0.7650	0.5731
% change in per-capita personal income	–1.7355	5.0205	–2.5864	5.9378
Unemp rate × mos not employed	0.0155**	0.0079	–0.0101	0.0167
Unemp rate × not employed last 12 mos	0.0265	0.1326	0.2078	0.3222
Unemp rate × not currently employed	–0.1474	0.1303	–0.0605	0.1339
Recruiter density	–1129.144	1497.839	–3.7306***	1.3491
Missing state	—	—	0.9887	6.3854
Expects more education	–0.3313	0.2328	0.5190	0.3953
Missing expects more education	–1.8358**	0.8229	–0.9339*	0.5110
Plans to get married in next 5 years	0.5385**	0.2413	0.1696	0.4257
Plans never to marry	–0.0461	0.3622	0.3685	0.5296
Ever been married	—	—	0.5247*	0.2984
Has children	—	—	–0.9775***	0.2674
Missing marital information	–0.3951	0.2483	–0.1382	0.3166
Months since school (natural log)	—	—	0.0507	0.1062
Some post-HS school	—	—	–0.4289	0.2949
Constant	–0.2491	4.5617	–7.8287	6.5827
Scale factor	0.0028		0.0188	
Log-likelihood	–1248.06		–562.67	
Chi-square	179.07		343.47	
Pseudo R-square	0.0726		0.3861	
Number of observations	7,621		3,554	

NOTES: Significance levels: ***0.01 level, ** 0.05 level, * 0.10 level. A "—" indicates that the variable was not included in the model.

than a model with just an intercept, but the low "pseudo R-square" shows that the model nevertheless explains only 7 percent of the variance for seniors but 38 percent of the variance for graduates.

A customary way to evaluate goodness of fit for a logit model is to create a table of "hits and misses." These tables designate observations with a predicted probability greater than 0.50 as a hit and those with a lower predicted probability as a miss (see Greene (1990), for example). This is not tractable for this problem, however, because enlistment is such a rare event in the NELS sample that all predicted probabilities were below 0.50. For this reason, we evaluate the fit of the model in a slightly different way in Table 5.2. This table shows the mean predicted probability of enlisting by actual enlistment status to evaluate whether the model is able to discriminate between individuals who do and do not enlist. The senior column reports very small differences between the predicted enlistment probabilities of enlistees and nonenlistees, further reinforcing the notion that this model has relatively low predictive power. The graduate model is better able to distinguish between enlistees and nonenlistees, as shown by the much larger predicted probability of enlisting for the graduate enlistees. Again this is consistent with the earlier information reported in Table 5.1.[2]

Now we compare our results to those of Hosek and Peterson. First, we compare the coefficient estimates, which are presented in Table 5.3 for seniors and Table 5.4 for graduates. In these tables, we compare the NELS estimates and the Hosek and Peterson estimates by computing a test statistic for significant differences between the two coefficients that approximates a t-statistic. Appendix B explains the test in detail. We present the approximate t-statistic along with the confidence level at which the two coefficients are statistically significantly different.

Table 5.2

Mean Predicted Probability of Enlisting for Seniors and Graduates, by Enlistment Status

Enlistment Status	Seniors	Graduates
Enlistees	0.0700	0.3933
Nonenlistees	0.0415	0.0468
Total	0.0428	0.0717

[2]Note that Hosek and Peterson (1985, 1990) do not include similar goodness-of-fit measures to which we can compare our results.

The first table, Table 5.3, shows that roughly a quarter of the senior coefficient estimates are different from the Hosek and Peterson estimates at the 95 percent significance level or higher. The small number of significantly different coefficient estimates may be related to the fact that many of the NELS coefficients were imprecisely estimated—that is, they have very wide standard errors that are likely to encompass the Hosek and Peterson estimate.

The most notable differences include some of the coefficients that were not statistically significant in our logit estimates, but that Hosek and Peterson did find significant. One of these is the coefficient on the black race indicator. Hosek and Peterson estimated that black men were significantly more likely to enlist than white men. Our coefficient estimate on this variable is negative, but not significantly different from zero. This finding is consistent with evidence reported in another portion of this project: Orvis et al. (1996) find declining black propensity rates. Also, we do not find that individuals' age when a senior affected their enlistment probability, while Hosek and Peterson found higher enlistment probabilities for older seniors and lower enlistment probabilities for younger seniors. We also found no influence on enlistment from per-capita personal income, in contrast to the large negative effect estimated by Hosek and Peterson. As discussed earlier, this is likely to stem in part from the fact that we use per-capita income at the state level, whereas Hosek and Peterson use county-level per-capita income.

Several other coefficient estimates differ significantly from the Hosek and Peterson estimates: mother's schooling, family income, and the unemployment rate. While Hosek and Peterson found that higher mother's schooling raised the probability of enlisting, we found that higher mother's schooling lowers enlistment probability. We also estimated a somewhat different relationship between enlistment and unemployment variables. Hosek and Peterson find that for graduates, individuals who are not employed are initially less likely to enlist, but after two months of being unemployed, their probability of enlistment is higher. In contrast, our results indicate the graduates who are not employed are initially more likely to enlist, but after one month of unemployment, they become less likely to enlist. Finally, we estimated different values for marital-expectations variables. For the variable "Plans never to marry," we did not estimate a significant

Table 5.3

Comparing NELS and Hosek and Peterson (H&P) Coefficient Estimates, Seniors Model

Variable	NELS Coefficient	H&P Coefficient	t-statistic for Difference	Confidence Level for Difference
Black	–0.4597	0.615	2.8306	0.99
Hispanic	–0.6047*	–0.023	1.3121	0.80
Age 17 when senior	0.0726	–0.444	–2.3447	0.95
Age 19+ when senior	–0.3507	0.504	2.2145	0.95
AFQT score (31–99)	–0.0134*	–0.010	0.3846	<0.50
Category IV indicator	–1.0251**	–0.948	0.1470	<0.50
AFQT score missing	–1.0022**	–0.182	1.4511	0.80
Mother's years of schooling	–0.1351***	0.101	4.3700	0.99
Mother worked	0.2744**	0.715	1.6421	0.90
Family income (in $ thousands)	–0.0186**	–0.030	–1.0827	0.70
Family income < $5,200	–0.2190	0.191	0.7645	<0.50
Family income missing	–1.0522**	–0.452	1.1400	0.70
Number of siblings	0.1268**	0.262	1.9281	0.90
Missing number of siblings	0.3306	—	—	—
Lives at home	–0.1202	0.249	0.9493	0.60
Missing lives at home	0.7312	—	—	—
Hourly wage (natural log)	–0.6430	–0.204	0.4494	<0.50
Wage < $2.25/hour	–1.2128*	–1.310	–0.0980	<0.50
Hourly wage missing	–1.1772	–0.713	0.2984	<0.50
Weekly hours currently employed	0.0135	0.044	1.8440	0.90
Months employed (natural log)	–0.1165	–0.244	–1.1128	0.70
Not currently employed	–0.1376	–0.414	–0.1446	<0.50
Weekly hours not currently employed	0.0127	0.009	–0.2821	<0.50
Months not employed	–0.1051*	0.010	0.9045	0.60
Not employed in last 12 months	–0.3204	0.417	0.6048	<0.50
Unemployment rate in county	0.1606*	–0.038	–2.0765	0.95
Share of seniors and recent graduates	2.3420	–0.076	–1.3582	—
Percent of population that is Hispanic	5.3661	–0.0009	–0.7645	0.50
Percent of population that is black	–0.0690	–0.034	0.0281	<0.50
Percent of labor force that is female	–0.0213	0.075	0.8782	0.60

Table 5.3 (continued)

Variable	NELS Coefficient	H&P Coefficient	t-statistic for Difference	Confidence Level for Difference
Per-capita personal income	–0.0197	–0.163	–3.2338	0.99
% change in per-capita personal income	–1.7355	–0.091	0.3275	<0.50
Unemp rate × not currently employed	0.0155**	–0.16	–1.3796	0.80
Unemp rate × not employed last 12 mos	0.0265	0.061	0.2336	<0.50
Unemp rate × mos not employed	–0.1474	0.037	1.3995	0.80
Recruiter density	–1129.144	1497.839	0.7543	0.60
Expects more education	–0.3313	–0.159	0.6431	<0.50
Missing expects more education	–1.8358**	—	—	—
Plans to get married in next 5 years	0.5385**	1.45	3.2797	0.99
Plans never to marry	–0.0461	3.07	4.5980	0.99
Missing marital information	–0.3951		1.5910	0.80
Constant	–0.2491	–4.12	–0.8130	0.60

NOTES: A "—" indicates that the variable was not included in the model. Asterisks indicate significance levels of logit estimates in Table 5.1. ***0.01 level, **0.05 level, *0.10 level.

coefficient, whereas Hosek and Peterson estimated a large positive coefficient. For the variable "Plans to get married in next 5 years," we estimated a value about one-third the size of the Hosek and Peterson estimate.[3]

We compare the coefficient estimates in our graduate model to those in Hosek and Peterson in Table 5.4. For graduates, many more coefficient estimates are significantly different from the Hosek and Peterson estimates at the 95 percent level or higher. Most of the differing coefficients are those related to labor force alternatives. Note that these are the same variables that were more often significant in the

[3]We do not compare coefficients on the "Share of seniors and recent graduates" variable, as this is defined differently in the two samples. See the variable descriptions in Appendix A, Table A.1, for more on this.

Table 5.4

Comparing NELS and Hosek and Peterson (H&P) Coefficient Estimates, Graduates Model

Variable	NELS Coefficient	H&P Coefficient	t-statistic for Difference	Confidence Level for Difference
Black	0.2200	0.878	1.9446	0.90
Hispanic	0.2396	–0.183	–0.9347	0.60
Age 16 when senior	–0.9093	—	—	—
Age 17 when senior	0.0493	–0.134	–0.7336	0.50
Age 19+ when senior	–0.4089	–0.258	0.3388	<0.50
AFQT score (31–99)	0.0117	0.0001	–1.1508	0.70
Category IV indicator	0.2919	–0.071	–0.5764	<0.50
AFQT score missing	0.7478	0.028	–1.1194	0.70
GED	–0.1084	0.4645	0.6580	<0.50
Mother's years of schooling	0.0280	0.074	0.5788	<0.50
Mother worked	0.1877	0.259	0.1704	<0.50
Family income (in $ thousands)	–0.0154*	–0.001	1.0048	0.70
Family income < $5,200	–1.5747***	–0.800	1.6899	0.80
Family income missing	–1.0555***	–0.223	1.9635	0.95
Number of siblings	0.0938	0.149	0.7138	0.50
Missing number of siblings	0.0266	—	—	—
Lives at home	–1.7779***	0.103	5.6988	0.99
Hourly wage (natural log)	–0.0469	–1.05	–4.2782	0.99
Hourly wage missing	4.2439***	–1.15	–6.5078	0.99
Weekly hours currently employed	0.0722***	–0.012	–6.8436	0.99
Months employed (natural log)	–0.0722***	–0.124	–0.3066	<0.50
Not currently employed	3.1304***	–3.09	–4.2938	0.99
Weekly hours not currently employed	0.0632***	0.029	–2.0412	0.99
Missing weekly hrs not currently emp	4.2066***	—	—	—
Months not employed	–0.3601***	0.040	2.8115	0.99
Months not employed in last 12 months	4.9063***	–1.40	–3.0396	0.99
Unemployment rate in county	0.0860	–0.048	–1.1256	0.70
Share of seniors and recent graduates	–1.6950	–0.234	0.5021	<0.50
Percent of population that is black	1.2234	–0.021	–0.9570	0.60
Percent of population that is Hispanic	–6.6640	0.008	0.7018	0.50

Table 5.4 (continued)

Variable	NELS Coefficient	H&P Coefficient	t-statistic for Difference	Confidence Level for Difference
Percent of labor force that is female	0.0060	0.041	0.2509	<0.50
Per-capita personal income	0.7650	–0.156	–1.6043	0.80
% change in per-capita personal income	–2.5864	–0.012	0.4336	<0.50
Unemp rate × not currently employed	–0.0605	–0.103	–1.0712	0.70
Unemp rate × not employed last 12 mos	0.2078	0.263	0.1644	<0.50
Unemp rate × mos not employed	–0.0101	0.070	0.9642	0.60
Recruiter density	–3.7306***	0.924	2.8464	0.99
Missing state	0.9887	—	—	—
Expects more education	0.5190	1.05	1.2969	0.80
Missing expects more education	–0.9339*	—	—	—
Plans to get married in next 5 years	0.1696	0.848	1.5218	0.80
Plans never to marry	0.3685	2.00	2.5468	0.99
Ever been married	0.5247*	0.737	0.5778	<0.50
Has children	–0.9775***	–0.059	2.6190	0.99
Missing marital information	–0.1382	—	—	—
Months since school (natural log)	0.0507	–0.392	–3.6115	0.99
Some post-HS school	–0.4289	–0.442	–0.0394	<0.50
Constant	–7.8287	1.75	1.4357	0.80

NOTES: A "—" indicates that the variable was not included in the model. Asterisks indicate significance levels of logit estimates in Table 5.1. ***0.01 level, **0.05 level, *0.10 level.

NELS estimates, and therefore have narrower standard error estimates. We also estimate significantly different values for two of the family-formation variables.

We find a significantly different estimate for the recruiter-density variable. While Hosek and Peterson find higher recruiter density is associated with greater enlistment probability, we estimate a negative and large coefficient on this variable. Several factors may contribute to the wide disparity in the estimates for this variable. First, our geographic unit differs from that available in Hosek and Peterson—we know only the state in which NELS respondents reside, while Hosek and Peterson knew the county. Our variable is therefore defined at the state rather than the county level. Second, recruiter density is a variable at the discretion of the Department of Defense and thus may vary with the enlistment probability and characteristics of the state or county. For example, in states where enlistment probabilities are low, more recruiters may be assigned to obtain a given number of recruits. Such optimizing decisions by recruiting managers may yield the seemingly anomalous finding that more recruiters are associated with *fewer* enlistments rather than more, as might be expected.

Next we examine differences between levels of explanatory variables used for the NELS estimates and the Hosek and Peterson estimates. As in our comparison of the coefficients, we compute an approximate t-statistic to test whether the NELS means are significantly different from the Hosek and Peterson means. Appendix B also describes this test. These results are reported in Table 5.5 for seniors and Table 5.6 for graduates.

In contrast to the comparisons for the coefficients, nearly every mean in the NELS senior model is significantly different from its counterpart in the Hosek and Peterson sample. Based simply on this table, it is difficult to generalize as to whether the changes in the regressors raise or lower the overall average likelihood of enlisting: among the regressors that statistically influence enlistment in the NELS estimates, some changed in ways that would be expected to raise enlistment and others changed in ways that would be expected to lower enlistment. For example, mean AFQT and mean months not employed changed in ways that would raise enlistment probability, according to the model. That is, mean AFQT score and months not employed are both lower in the NELS than in the Hosek and Peterson data, and higher levels of both of these variables are associated with lower enlistment probability. On the other hand, other significant variables changed in ways that would be expected to lower enlistment probabilities, including the fraction of Hispanics, mother's schooling, family income, the number of siblings, and the number of individuals planning to wed in the next five years. The simulation reported below will estimate the net effect of all of these changes in the levels of regressors in the model.

Table 5.5

Comparing NELS and Hosek and Peterson (H&P) Regressor Means, Seniors Model

Variable	NELS Mean	H&P Mean	t-statistic for Difference	Confidence Level for Difference
Black	0.1067	0.1173	0.6984	0.50
Hispanic	0.0944	0.0496	–4.2010	0.99
Age 17 when senior	0.4192	0.5184	4.0690	0.99
Age 19+ when senior	0.1148	0.0565	–5.1466	0.99
AFQT score (31–99)	61.4393	66.0702	4.9413	0.99
Category IV indicator	0.2177	0.2452	1.3180	0.80
AFQT score missing	0.2344	0.0706	–12.6926	0.99
Mother's years of schooling	13.3061	12.0265	–10.4267	0.99
Mother worked	0.8523	0.5145	–14.1197	0.99
Family income (in $ thousands)	21.4166	25.2791	5.7332	0.99
Family income < $5,200	0.0609	0.0268	–4.3031	0.99
Family income missing	0.1798	0.1729	–0.3712	<0.50
Number of siblings	1.9085	3.0786	12.1119	0.99
Missing number of siblings	0.1176	—	—	—
Missing lives at home	0.1462	—	—	—
Hourly wage (natural log)	0.4899	1.1518	66.2011	0.99
Wage < $2.25/hour	0.0217	0.1008	5.4182	0.99
Hourly wage missing	0.0759	0.2088	6.4944	0.99
Weekly hours currently employed	8.5683	18.8080	15.6472	0.99
Months employed (natural log)	0.945	1.7746	12.9096	0.99
Not currently employed	0.4865	0.2481	–11.1807	0.99
Weekly hours not currently employed	3.3123	30.0294	39.9992	0.99
Lives at home	0.7828	0.9477	2.8393	0.99
Months not employed	2.6926	5.1252	16.4231	0.99
Not employed in last 12 months	0.1757	0.1166	–3.7940	0.99
Unemployment rate in county	6.7587	6.0278	–7.8077	0.99
Share of seniors and recent graduates	0.4438	0.1484	—	—
Percent of population that is black	0.1217	0.1157	–1.0072	0.60
Percent of population that is Hispanic	0.0219	0.0575	7.9810	0.99

Table 5.5 (continued)

Variable	NELS Mean	H&P Mean	t-statistic for Difference	Confidence Level for Difference
Percent of labor force that is female	45.6189	41.4478	–28.4887	0.99
Per-capita personal income	7.7789	8.7681	8.9329	0.99
% change in per-capita personal income	0.0444	0.3785	3.3804	0.99
Unemp rate × not currently employed	3.3316	1.4707	–13.8279	0.99
Unemp rate × not employed last 12 mos	1.2156	0.7236	–4.7303	0.99
Unemp rate × mos not employed	18.2548	7.4308	–12.1615	0.99
Recruiter density	0.0791	0.0005	–86.9162	0.99
Expects more education	0.9044	0.6211	–12.1970	0.99
Missing expects more education	0.0386	—	—	—
Plans to get married in next 5 years	0.0846	0.3885	13.4025	0.99
Plans never to marry	0.0467	0.0255	–2.8766	0.99
Missing marital information	0.1846	—	—	—
Constant	1.000	1.0000	—	—

NOTE: A "—" indicates that the variable was not included in the model.

We compare the regressors for the graduate model in Table 5.6. For the graduate regressors, every mean is different at the 95 percent level or more except for one, the fraction ever married. As was the case for the senior variables, the means of some variables changed in a direction that would lead to higher enlistment rates and others changed in a direction that would tend to lower enlistment rates. Simply inspecting these means without factoring in the size of their effects on enlistment probability—as identified by coefficient values—makes it difficult to generalize the effect of changes in these means on overall enlistment probabilities. We next present a simulation that addresses this issue.

Table 5.6

Comparing NELS and Hosek and Peterson (H&P) Regressor Means, Graduates Model

Variable	NELS Mean	H&P Mean	t-statistic for Difference	Confidence Level for Difference
Black	0.1730	0.0978	–6.5913	0.99
Hispanic	0.1280	0.0416	–9.9543	0.99
Age 16 when senior	0.0040	—	—	—
Age 17 when senior	0.3176	0.3888	3.8514	0.99
Age 19+ when senior	0.2082	0.0808	–12.5984	0.99
AFQT score (31–99)	53.53193	64.3344	15.9100	0.99
AFQT score missing	0.3893	0.1080	–20.9830	0.99
Category IV indicator	0.2861	0.2178	–4.2683	0.99
GED	0.1208	0.0449	–8.5753	0.99
Mother's years of schooling	12.4844	11.6826	–9.1048	0.99
Mother worked	0.7701	0.5013	–14.5754	0.99
Family income (in $ thousands)	14.3511	21.7281	14.4119	0.99
Family income < $5,200	0.1170	0.0640	–5.2900	0.99
Family income missing	0.1855	0.1576	–1.9829	0.95
Number of siblings	2.0747	3.2512	15.2977	0.99
Missing number of siblings	0.1746	—	—	—
Lives at home	0.5480	0.6900	7.9111	0.99
Hourly wage (natural log)	0.6836	1.5098	47.2622	0.99
Hourly wage missing	0.0289	0.1254	8.1091	0.99
Weekly hours currently employed	28.4866	42.2065	27.3791	0.99
Months employed (natural log)	1.8032	2.2080	8.2724	0.99
Not currently employed	0.2735	0.1129	–12.4360	0.99
Weekly hours not currently employed	7.4237	37.5300	62.9000	0.99
Missing weekly hrs not currently emp	0.0876	—	—	—
Months not employed	2.2621	2.8990	4.7949	0.99
Not employed in last 12 months	0.0852	0.0287	–8.2578	0.99
Unemployment rate in county	6.2505	6.0557	–2.5314	0.99
Share of seniors and recent graduates	0.4421	0.1507	—	—
Percent of population that is black	0.1210	0.1091	–2.5700	0.99
Percent of population that is Hispanic	0.0207	0.0493	10.0631	0.99
Percent of labor force that is female	42.0591	41.5396	–2.2538	0.99

Table 5.6 (continued)

Variable	NELS Mean	H&P Mean	t-statistic for Difference	Confidence Level for Difference
Per-capita personal income	9.7384	8.7718	–12.6868	0.99
% change in per-capita personal income	0.0433	0.1997	1.9879	0.95
Unemp rate × not currently employed	13.4036	0.7167	–18.9718	0.99
Unemp rate × not employed last 12 mos	0.5029	0.1917	–6.5800	0.99
Unemp rate × mos not employed	1.7278	2.2179	1.7156	0.90
Recruiter density	0.0828	0.0005	–46.8616	0.99
Missing state	0.0783		–17.3762	0.99
Expects more education	0.6185	0.4300	–10.1036	0.99
Missing expects more education	0.2793	—	—	—
Plans to get married in next 5 years	0.0804	0.5494	27.2171	0.99
Plans never to marry	0.0422	0.0288	–2.1267	0.99
Ever been married	0.1150	0.1299	1.1684	0.70
Has children	0.1660	0.0835	–7.2558	0.99
Missing marital information	0.3686	—	—	—
Months since school (natural log)	0.7724	2.8256	60.3703	0.99
Some post-HS school	0.2471	0.1453	–7.2398	0.99
Constant	1.000	1.0000	—	—

NOTE: A "—" indicates that the variable was not included in the model.

The final dimension of our comparison of the NELS estimates and the Hosek and Peterson estimates is a simple simulation that assesses whether the relative impact of changes in enlistment probabilities estimated by the two are due more to changes in coefficients or changes in characteristics over the period. To explain the simulation we conduct, we let the function $F(\cdot)$ represent the logistic regression function, as in

$$\Pr(Y=1)=\frac{e^{\beta_n' X_m}}{1+e^{\beta_n' X_m}}=F(\beta_n' X_m),$$

where n and m are one when they refer to the Hosek and Peterson coefficients or regressors, respectively, and are two when they refer to the NELS coefficients and regressors.

The difference between the predicted enlistment rate using the Hosek and Peterson coefficient estimates and regressors, $F(\hat{\beta}_1' X_1)$, and the predicted enlistment rate using the NELS coefficient estimates and regressors, $F(\hat{\beta}_2' X_2)$ can be decomposed into a portion due to differences in the regressors and a portion due to differences in the coefficients. In equation form, this is

$$F(\hat{\beta}_2' X_2) - F(\hat{\beta}_1' X_1) = \left[F(\hat{\beta}_1' X_2) - F(\hat{\beta}_1' X_1)\right] + \left[F(\hat{\beta}_2' X_2) - F(\hat{\beta}_1' X_2)\right]. \quad (2)$$

The first term in brackets on the right-hand side yields the difference in the predicted probabilities due to changes in the regressors. This term shows the difference in predicted enlistment between using the two different sets of regressors, X_2 and X_1, when using the Hosek and Peterson coefficient estimates, β_1. This expression keeps behavior the same—that is, the coefficients in the model stay the same—but alters the characteristics of individuals in the pool of potential enlistees—that is, alters the characteristics. The second term in brackets yields the difference in predicted probability due to using different estimated coefficients, $\hat{\beta}_2$ and $\hat{\beta}_1$, when using the NELS regressors, X_2. This expression holds the types of individuals in the sample constant—the X's stay the same—but allows their behavior to change—that is, allows the $\hat{\beta}$ to differ. In this way the total expression parcels out net changes between the NELS and Hosek and Peterson estimates into those due to changing behavioral responses and those due to the types of individuals at risk to enlist.

The logistic regression model has the property that the mean predicted probability will equal the actual probability in the sample (see Greene, 1990). Hence, in equation (2) we can substitute the Hosek and Peterson sample enlistment rate for $F(\hat{\beta}_1 X_1)$ and the NELS sample enlistment rate for $F(\hat{\beta}_2' X_2)$. The only other term in equation (2) is $F(\hat{\beta}_1' X_2)$. We estimate this by predicting enlistment probabilities for each member of the NELS sample using the Hosek and Peterson coefficients and then taking the mean.

Using these methods for obtaining each of the predicted enlistment rates, we can rewrite equation (2) for seniors as

$$F(\hat{\beta}_2'X_2)-F(\hat{\beta}_1'X_1)=\left[F(\hat{\beta}_1'X_2)-F(\hat{\beta}_1'X_1)\right]+\left[F(\hat{\beta}_2'X_2)-F(\hat{\beta}_1'X_2)\right]$$

$$.043 \quad - \quad .039 \quad = \left[\quad .278 \quad - \quad .039 \quad\right] + \left[\quad .043 \quad - \quad .278 \quad\right]$$

$$.004 \qquad = \qquad .239 \qquad + \qquad (-.235).$$

Several important points come out of this expression.[4] First, the enlistment rate predicted for the NELS sample using the Hosek and Peterson coefficients, $F(\hat{\beta}_1'X_2)$, is 0.278, which is higher than the enlistment rate predicted for the NELS sample using the NELS coefficients (0.043). This indicates that on the whole, the changes in the means we observed in Table 5.5 are in the direction of individuals being more likely to enlist on average than was the case in the Hosek and Peterson sample. This same point is made by the fact that the first term in the right-hand side of the equation above, the change in enlistment rates due to changes in regressors, is positive 0.239. This says that the NELS regressors lead to higher average enlistment rates based on the Hosek and Peterson model than the Hosek and Peterson regressors.

The second term in the expression shows the net change in enlistment rates due to differences between the NELS and the Hosek and Peterson coefficients. This value is estimated at –0.235. This shows that changes in coefficients went the opposite way over the period—that NELS coefficients would predict lower enlistment rates for the NELS sample than would the Hosek and Peterson coefficients. This has the implication that had we used the Hosek and Peterson model to generate predictions for the NELS sample, we would have greatly overstated its enlistment rates.

[4]Since the number of accessions in 1992 was lower than the number of accessions in 1980, it may seem surprising that the enlistment *rate* in 1992 is slightly higher than it was in 1980. However, the size of the 18-year-old male population was also lower in 1992 than in 1980 (see Klerman and Karoly, 1994).

Note that since enlistment rates are relatively similar in the two samples—0.043 for the NELS sample and 0.039 for the Hosek and Peterson sample—the fact that the regressors changed in a way that raised the probability of enlistment implies that the coefficients must have changed in a way to lower predicted enlistment rates.[5] Recall that the individual coefficients indicate the marginal effect of a given characteristic on the probability of enlisting. This change in coefficients could represent a change in youth behavior: on average, youths in 1992 were less likely to enlist than youths in 1980. The change in coefficients could also represent a change on the part of the military: the military could have been less likely to admit youths in 1992 than in 1980.

For graduates, the expression is

$$
\begin{aligned}
F(\hat{\beta}_2' X_2) - F(\hat{\beta}_1' X_1) &= \left[F(\hat{\beta}_1' X_2) - F(\hat{\beta}_1' X_1)\right] + \left[F(\hat{\beta}_2' X_2) - F(\hat{\beta}_1' X_2)\right] \\
.072 \quad - \quad .053 &= \left[\ .003 \quad - \quad .053\ \right] + \left[\ .072 \quad - \quad .003\ \right] \\
.019 &= \quad (-.050) \quad + \quad .069.
\end{aligned}
$$

The story here is the opposite of that for the senior model. First, we find that predicting NELS graduate enlistments using the Hosek and Peterson coefficients, $F(\hat{\beta}_1' X_2)$, would have led to a predicted rate of enlistment close to zero, 0.003. Along with the negative estimate for the first term of the expression, –.050, this indicates that the effect of changes due to regressors over the period was to lower predicted enlistment. That is, more graduates in the NELS sample than in the Hosek and Peterson sample had characteristics that the Hosek and Peterson model suggests would make them less likely to enlist. The estimate of 0.69 for the second term on the right-hand side of the expression indicates that the coefficients in the NELS senior model lead to higher predicted enlistment rates than the coefficients estimated by Hosek and Peterson.

---

[5]It would be interesting to assess whether there is an analogous relationship for aggregate models of enlistment.

NEW BIVARIATE SPECIFICATION

Next we present results from a bivariate enlistment model that adds some new variables and omits some variables from the Hosek and Peterson specification. We add several variables to capture some additional factors that the theoretical discussion above suggests may be important to the enlistment decision. We are particularly interested in capturing factors that may have been relatively unimportant at the time of the Hosek and Peterson studies but have emerged in more recent years as potentially important. One such variable is the average in-state tuition at a four-year institution in the respondent's state. This identifies individuals who face higher in-state college tuition costs and would be more likely to enlist as a result.

Other new variables are as follows. We add a variable that indicates whether a youth was from an immigrant household or not. As discussed above, immigration has emerged as one of the most important issues relating to youth demographics in the 1990s. To proxy for being in an immigrant household, our variable indicates when English is not an individual's first language. It is unclear whether this variable would be associated with higher or lower enlistment probability. We also add a variable to show when an individual has a parent serving in the military in 1992. We would expect youths from families with strong military traditions to be more likely to enlist.

In addition, we add two variables to indicate whether the individual is likely to meet the military's moral standards. One of these variables indicates whether a youth reported ever using marijuana, and the other indicates whether the youth or one of his friends had ever been arrested. In addition, all of these added variables may have a companion variable to indicate when the value is missing. The precise definitions of these variables are presented in Table A.1 and their means are in Table A.2, both of them in Appendix A.

We also delete some variables from the specification above. We omit variables primarily because there is a possibility that they may be endogenous to the enlistment or other post-high-school activity decision—that is, it may be the case that the value the variable takes on is a result of the activity decision. For instance, one such variable is the indicator of whether the respondent lives at home. If the respondent is a senior and plans to attend college away from home or enlist in the military, it is unlikely that this person will incur the costs of setting up a separate household knowing that he will be relocating at the end of his senior year. In this case, we would expect the value of the variable to be one for such individuals. Here, the value of this variable is clearly correlated with the decision we are trying to estimate, violating one of the basic assumptions of multivariate regression and logit models—that the covariates be uncorrelated with the error term of the equation for the variable you are trying to estimate.

Another reason that we omit some variables in this specification is that their meaning for the populations in our study is unclear. This is especially true of the labor-market variables for graduates who have enlisted. For example, when asked if they are currently employed, graduates who are in the Delayed Entry Program (DEP) are likely to answer "no," as many individuals take a break between school or a job and beginning their military training. In fact, we find that a disproportionate share of graduate enlistees report that they are not currently employed. It is unclear whether not being employed induced them to enlist or whether having signed up to join the military induced them to not be employed. In this specification we also omit the other variables related to the individual's labor-market participation. Similarly, we do not include the variables that indicate time since postsecondary education and whether the respondent has some postsecondary education.

We present estimates for this new specification in Table 5.7. The first two columns contain estimates for seniors, and the second two columns contain estimates for graduates. As in our replication of the Hosek and Peterson model, presented in Table 5.1, here we also find that only a modest number of variables are significant. Among the most notable findings for seniors is the negative coefficient for the black indicator variable. This is directly opposite the findings in Hosek and Peterson and counter to conventional wisdom that blacks are overrepresented in the enlisted force and are more likely to enlist. It is well known, however, that blacks often have other characteristics that make them more likely to enlist, such as lower family income and lower mother's education, which could still result in blacks' overrepresentation despite the negative coefficient on the black indicator variable. This finding is consistent with other results from this project reported in Orvis et al. (1996), which finds a declining enlistment propensity for blacks in the 1990s.

Table 5.7

Logistic Regression Estimates of Enlistment Probability Using New Specification, for Seniors and Graduates

	Seniors		Graduates	
Variable	Coefficient Estimate	Standard Error	Coefficient Estimate	Standard Error
Black	–0.6067**	0.3040	0.1981	0.2664
Hispanic	–0.2765	0.3470	0.3334	0.3267
Age 16 when senior	—	—	–0.2033	1.0399
Age 17 when senior	0.0575	0.1782	–0.0616	0.1977
Age 19+ when senior	–0.1706	0.2982	–0.4251	0.3266
AFQT CAT I indicator	–1.3110**	0.6778	1.7400***	0.6695
AFQT CAT II indicator	–0.2065	0.3533	0.9110**	0.4313
AFQT CAT IIIB indicator	0.2654	0.2604	0.6044	0.4212
AFQT CAT IV indicator	–0.2649	0.2864	0.0224	0.4014
AFQT CAT V indicator	–0.1553	0.4178	–1.5263**	0.6614
AFQT score missing	–0.2715	0.3032	0.6472	0.4296
GED	—	—	0.1559	0.5056
Mother's years of schooling	–0.1299***	0.0360	–0.0039	0.0495
Mother worked	0.3948*	0.2153	0.2228	0.2959
Family income (in $ thousands)	–0.0215**	0.0090	0.0004	0.0061
Family income < $5,200	–0.1966	0.3679	–0.7002**	0.3575
Family income missing	–0.4142	0.2869	–0.4854	0.3042
Number of siblings	0.1361**	0.0555	0.0622	0.0588
Missing number of siblings	0.3433	0.3035	–0.0981	0.3901
Unemployment rate in county	0.1342**	0.0682	0.0674	0.0941
Percent of population that is black	–0.8179	1.1015	2.5226**	1.0913
Percent of population that is Hispanic	–2.9952	6.5836	4.4664	7.0593
Percent of labor force that is female	–0.0283	0.1007	0.0805	0.1221
Per-capita personal income	–0.0192	0.0224	–0.3977	1.1198
% change in per-capita personal income	1.7974	4.9344	–2.9325	6.3858
Recruiter density	–0.6256	1.3506	1.3160	1.0936
Missing state	—	—	3.6780	5.5072
Expects more education	–0.4721**	0.2320	0.2037	0.3759
Missing expects more education	–2.2871***	0.7515	–0.0450	0.8178
Plans to get married in next 5 years	0.4136*	0.2448	0.0577	0.3682
Plans never to marry	–0.1461	0.3623	0.2824	0.4094
Ever been married	–0.2687	0.4797	0.8453***	0.3042
Has children	0.5488	0.4546	–0.4715	0.3596
Missing marital information	–0.4915*	0.2955	–0.0354	0.3546

Table 5.7 (continued)

Variable	Seniors Coefficient Estimate	Seniors Standard Error	Graduates Coefficient Estimate	Graduates Standard Error
Parent in the military	0.6452	0.3994	1.6914***	0.5405
Missing parent in military	0.3589	0.2549	–0.4812	0.3352
English not first language	–1.0381***	0.3785	–0.2646	0.3649
Missing English language info	–0.1146	0.4576	0.2859	0.5334
Uses marijuana	0.1011	0.1703	–0.5474	0.2185
Missing marijuana use	0.0141	0.3394	–0.6700**	0.3883
R or friend has been arrested	0.4951	0.3823	0.1151*	0.4907
Missing arrest info	0.7344	0.7393	–0.7740	0.6688
Average in-state tuition	0.0000	0.0001	0.0001	0.0002
Constant	–0.3106	4.6998	–7.3424	5.5179
Scale factor	0.0312		0.0430	
Log-likelihood	–1301.20		–804.73	
Chi-square	161.37		185.32	
Pseudo R-square	0.0754		0.1459	
Number of observations	7,953		3,798	

NOTES: A "—" indicates that the variable was not included in the model. Significance levels: *** 0.01 level, ** 0.05 level, * 0.10 level.

In contrast to the earlier estimates, in this specification we find positive effects on enlistment probability of the mother worked variable. We also now find that enlistment probability is lower for individuals who expect more education.

In terms of the innovations to the Hosek and Peterson specification, we find only a few significant effects. We observe that when using AFQT categories rather than continuous AFQT score, the only significant effect is that individuals in the CAT I group are substantially less likely to enlist. This suggests that the overall negative coefficient estimated earlier for the continuous AFQT score is likely to be driven primarily by the low enlistment rates of individuals at the very upper portions of the AFQT distribution. Among the variables added in this specification, only not having English as a first language is significant. This leads to much lower rates of enlistment.

In the new graduate specification, we also estimate only a modest number of significant variables. There are also a few new findings in this specification for graduates. In contrast to seniors, we estimate that individuals in the upper AFQT categories are more rather than less likely to enlist. Among the variables added to this specification, we find that different types of variables are significant. First, having a parent in the military raises the probability of enlistment. Second, we find some significant, albeit somewhat puzzling, results for the drug use and arrest variables. While marijuana use itself is not significant, individuals missing this variable are less likely to join the military. It is unclear how to interpret this effect, since a large number of individuals had missing values for this variable, presumably because of the sensitive nature of the question. We find that having been arrested or having a friend who has been arrested raises the likelihood of enlisting, which is surprising given that this variable was expected to proxy for having difficulty meeting the moral requirements for enlistment.

Examining the goodness of fit of this model in Table 5.8, we see that for seniors, this new specification is slightly better able to distinguish between enlistees and nonenlistees, but not much better. There is still only a very small difference between the mean predicted probability of those who enlist and those who do not. For graduates, this model discriminates between enlistees and nonenlistees less well than the Hosek and Peterson specification. This can be seen by the much lower predicted probability of enlistment in this table, 0.17, compared to that in Table 5.2, 0.39.

Table 5.8

Mean Predicted Probability of Enlisting for Seniors and Graduates, by Enlistment Status

Subsample	Seniors	Graduates
Enlistees	0.0724	0.1705
Nonenlistees	0.0415	0.0604
Total	0.0429	0.0679

ESTIMATING THE TRIVARIATE CHOICE MODEL

We estimate the trivariate specification using only data from the 1994 wave of the NELS, as it is only possible to ascertain whether the individual chose to attend college, work, or none of these activities after some time has elapsed since high school graduation. We define the three choices as follows. An individual is considered to have enlisted if he is counted as enlisted in either the senior model or graduate model above. Given that an individual is not in the enlisted group, he is eligible to be counted in one of the other two groups. An individual is in the *College* group if he was enrolled full-time in college, which could be two-year or four-year, for at least 12 months between June 1992 and August 1994. These months need not be contiguous. Individuals in neither the *Enlisted* nor the *College* group are eligible to be in the *Working/Other* group. A respondent was considered to be in the *Working* group if he was in the labor force for at least 12 months between June 1992 and August 1994 and his monthly salary was at least $500.00 in the months he was in the labor force. Individuals who fell into none of these categories were considered to be in the *Other* category. However, because such a small fraction of respondents were in this last category, and because their descriptive characteristics more closely aligned with the *Working* group, we combined them with the *Working* group to form the *Working/Other* group.

The fraction of respondents in each of these three categories is shown in Table 5.9. About equal numbers of individuals choose to attend college and work, while about 6.4 percent enlist. The marginal effects of a unit change in each regressor on the probability of making each choice are reported in Table 5.10, along with asterisks that indicate the significance level of the estimates. For estimation, the omitted category is *Enlisted*. The coefficient estimates of our trivariate logit model using the variables in the new specification are

Table 5.9

Percent Making Each Choice in the Trivariate Model

Enlist	Attend College	Work/Other
6.41	46.36	47.23

Table 5.10

Change in Probability of Making Each Choice for One Unit Change in Regressor

Variable	Enlist	Attend College	Work/ Other
Black	0.0116	–0.0653**	0.0537
Hispanic	0.0055	–0.0398	0.0343
Age 16 when senior	0.0042	–0.0079	0.0037
Age 17 when senior	0.0031	0.0054	–0.0085
Age 19+ when senior	–0.0104	–0.0137	0.0241
AFQT CAT I indicator	0.0225	0.2391	–0.2617*
AFQT CAT II indicator	0.0164	0.0454	–0.0618*
AFQT CAT IIIB indicator	0.0296	–0.0870***	0.0574*
AFQT CAT IV indicator	0.0037	–0.2095**	0.2059*
AFQT CAT V indicator	–0.0044	–0.2667	0.2711**
AFQT score missing	0.0326	–0.1301***	0.0975
GED	0.0232	–0.2372***	0.2141
Mother's ed: less than high school	0.0219	–0.1109***	0.0890
Mother's ed: some college	–0.0007	0.0780	–0.0773
Mother's ed: college degree	–0.0069	0.1876***	–0.1807
Mother's ed: postcollegiate	–0.0266	0.2098***	–0.1833
Missing mother's education	0.0375	–0.0783***	0.0408***
Mother worked	0.0138	0.0534	–0.0672**
Family income (in $ thousands)	–0.0005	0.0046***	–0.0041
Family income < $5,200	–0.0267	0.0247*	0.0021*
Family income missing	–0.0293	0.1056***	–0.0763*
Number of siblings	0.0086	–0.0336***	0.0250**
Missing number of siblings	0.0024	–0.0933	0.0909
Unemployment rate in county	0.0087	–0.0275***	0.0189
Percent of population that is black	0.0868	0.1302*	–0.2171***
Percent of population that is Hispanic	–0.1707	1.5238	–1.3531
Percent of labor force that is female	0.0018	–0.0009	–0.0010
Per-capita personal income	–0.0881	0.1235***	–0.0353**
% change in per-capita personal income	–0.0967	–1.2809	1.3776
Recruiter density	0.1310	–0.3982***	0.2671***
Missing state	0.0556	–0.2427	0.1871
Expects more education	–0.0606	0.5326***	–0.4720
Missing expects more education	–0.0847	0.4075***	–0.3228
Plans to get married in next 5 years	0.0310	–0.1924***	0.1614
Plans never to marry	0.0322	–0.1480***	0.1158
Ever been married	0.0825	–0.3919***	0.3094***
Has children	0.0029	–0.2336***	0.2307***
Missing marital information	–0.0122	–0.0154	0.0276
Parent in the military	0.0995	–0.2271***	0.1276***

Table 5.10 (continued)

Variable	Enlist	Attend College	Work/ Other
Missing parent in military	0.0117	–0.0901*	0.0784
English not first language	–0.0435	0.2019***	–0.1584
Missing English language info	–0.0329	0.0329***	–0.1861
Uses marijuana	–0.0077	–0.1135	0.1211**
Missing marijuana use	–0.0154	–0.1260	0.1414**
R or friend has been arrested	0.0145	–0.1178*	0.1033
Missing arrest info	–0.0133	–0.2918	0.3051
Average in-state tuition	0.0000[a]	0.0000[a]	0.0000[a]*
Constant	–0.1722	–0.1823	0.3545
Log-likelihood	–5204.76		
Chi-square	3791.10		
Pseudo R-square	0.2670		
Number of observations	8,009		

NOTES: A "—" indicates that the variable was not included in the model. Significance levels: *** 0.01 level, ** 0.05 level, * 0.10 level.

[a]The estimates of these coefficients are less than 0.00005.

reported in Appendix Table A.3. Note that these estimates are not directly comparable to the earlier estimates because this model is estimated on a slightly different sample that includes individuals in both the senior and graduate samples of the earlier estimates. For continuous variables in Table 5.10, the coefficient indicates the change in the probability of making each choice when the variable changes by one unit. For dummy variables, the coefficients show the change in the probability when the variable changes from zero to one.

Separating nonenlistment into the college and work/other alternatives better shows why a variable might be related to the probability of enlisting: by affecting college alternatives or work/other alternatives. For example, we found in the previous estimates that family income lowered the probability that seniors enlisted and raised the probability that graduates enlisted. Table 5.10 shows that this is likely to be because family income raises the likelihood of attending college, which is probably more relevant for seniors, and is negatively associated with the work/other choice, which is probably more salient for graduates.

Also, we observe many more significant effects in this specification than in earlier specifications. This is likely to be associated with the fact that in this specification, a variable can have two opposite effects on the two nonenlistment choices, whereas when these choices are lumped together in the nonenlistment category, the two effects could cancel each other to elicit an estimate of zero. An example of a variable that has contradictory effects on enlistment via the two other alternatives is AFQT category. We find that individuals with lower scores are less likely to attend college but are more likely to make the work/other choice. On net this yields small positive marginal effects of AFQT on enlistment.

In this specification, we find that blacks are slightly more likely to enlist, primarily because they are less likely than whites to attend college. We also find that the variables related to mother's education, family income, and number of siblings affect enlistment primarily through their effects on the probability of attending college. This is consistent with the theoretical underpinnings for including these variables in a model of enlistment probability—that they would influence enlistment through their relationship to college choice alternatives. Educational expectations also operate by affecting the college choice probability, as expected. Expectations on family formation also influence enlistment through the college choice, largely by reducing the probability of attending college. Having a mother who worked makes enlistment more likely, primarily because it makes work less likely. Each of these effects is consistent with the military being an attractive opportunity for seniors from less privileged backgrounds.

Among the new variables being explored in this report, we find results largely consistent with earlier findings. We estimate that not having English as a first language lowers enlistment probability primarily by substantially raising the likelihood of attending college. Marijuana use enters the model primarily by raising the likelihood of making the work/other choice, while arrest information enters primarily by reducing the chances of attending college. In-state tuition exerts a small effect in this model, surprisingly through raising the work/other choice likelihood.

Note that the signs of the estimated changes in probability for "attend college" and "work/other" are often opposite. This is probably due to the fact that many of the variables elicit a relatively small change in the probability of enlisting, meaning that the change in probability for enlisting is often not far from zero. Hence, the changes in probability for the other two choices must often offset each other.

The goodness-of-fit assessment presented in Table 5.11 shows that the mulitnomial model is better able to identify enlistees than the bivariate specifications estimated earlier. It assigns a much higher probability of enlisting to those who actually enlist than did the senior bivariate model but not as high as the graduate model. This specification does an even better job of discriminating between those who attend college and those who do not, and those who choose work/other and those who do not. It is interesting to note that enlistees are on average more likely to attend college than those who worked, and are much more likely to work than those who attended college. Also, the group that chose work/other has a higher mean enlistment probability than the college attendees.

Table 5.11

Mean Predicted Probability of Making Each Choice, by Actual Choice

	Predicted Probability Alternative		
Actual Choice	Enlist	Attend College	Work/ Other
Enlist	.1284	.3798	.4918
Attend college	.0539	.6568	.2893
Work/other	.0651	.2852	.6497
Total[a]	.0641	.4636	.4723

[a]This total predicted probability equals the actual enlistment probability, consistent with the properties of the logit model mentioned earlier.

Chapter Six

CONCLUSIONS

The primary value of the individual enlistment decision models estimated in this report is that they identify the characteristics of youths who are most likely to enlist. These results have the potential to increase recruiting efficiency by allowing recruiters to focus their efforts on those individuals who are most likely to enlist and by helping in the design of policies to attract those with a low probability of enlisting. Our replication of the Hosek and Peterson models, which were estimated using data from the early 1980s, generally finds that similar variables raised the likelihood of enlistment in the NELS data from the early 1990s. In other words, despite the many changes that took place between 1980 and 1992, the same characteristics were associated with enlistment decisions in the two periods. For seniors, we find that AFQT score, mother's schooling, family income, number of siblings, marital plans, and only a couple of work-related variables are important determinants of enlistment choice. For graduates, family income, number of siblings, and marital plans were still related to enlistment probability, but a great number of the work-related variables were also important, including wage-related variables, employment status, duration of unemployment, and other variables. These differences between seniors and graduates make sense because while many seniors' primary alternative activity is college, the graduates have already decided not to attend school and therefore their competing alternative is more likely to be the labor force.

In addition to identifying which characteristics are associated with youths' enlistment decisions, this report makes progress toward understanding more about the competition recruiting faces. That is, by using a three-choice model where the choices are enlisting, going to college, or working in the civilian sector, we show which activities youth are likely to choose instead of the military. This is important for designing recruiting incentives because it allows the military to tailor the incentives to make enlistment more attractive relative to the next-best alternative. For example, if college attendance is the next-best alternative, recruiting incentives might want to stress educational benefits or on-the-job training. But if civilian employment is the next-best alternative, recruiting incentives might focus on job security, wage comparability, or benefits. Many of the estimates of the trivariate model imply a high degree of substitutability between the college/military choice for high-quality youth. The results imply that competition for recruits among high-quality youth derives largely from higher education opportunities, so recruiting resources should be directed in a way that recognizes college as an important alternative activity. The results also point to the importance of civilian employment as the most important source of competition for potential recruits who are not college bound.

Overall, the trivariate model yields more insightful and plausible results on enlistment decisions. These results substantiate the theoretical underpinnings behind the individual enlistment decision model by specifying whether a characteristic is associated with enlistment because it raises or lowers the attractiveness of enlisting relative to other choices. By comparing enlistment to attending college and working, the results show much more clearly whether a characteristic that predicts enlistment does so because it makes recruiting more attractive relative to college or relative to work.

Furthermore, the three-choice model also demonstrates whether a particular variable influences enlistment via a substitution effect between the military/college choice or between the military/work choice. For example, we find that variables included in the model primarily because they are believed to influence the expected returns to education—such as AFQT score, age when a senior, and mother's education—lower enlistment rates because they raise the likelihood of attending college rather than by operating through the work/other choice. These findings suggest that youths considering college may respond well to recruiting incentives associated with attending college, such as college benefit programs. When contrasting the decision to enlist or attend college, there was no significant difference between the probability of choosing one or the other for youths in AFQT CAT I through AFQT CAT IIIA, but youths with lower test scores were less likely to choose to attend college. Hence the biggest differences between college attendees and high-quality recruits may not be test scores but rather the availability of resources to pay for college. However, when comparing enlisting and working, we found that individuals in the lower part of the AFQT distribution were most likely to enlist. For individuals who have chosen not to attend college, the middle part of the AFQT distribution might be the most fertile ground for recruiting efforts.

Generally, the characteristics that we found to be important to enlistment decisions were similar to those Hosek and Peterson found to be important. However, we did find a number of differences that supported our hypotheses on the likely effects of changes that took place between the two time periods, as enumerated in Chapter Three. For example, we hypothesized that the greater share of blacks in the 1992 youth population would be associated with a smaller coefficient on the black indicator variable. Earlier papers (Hosek and Peterson, 1985, 1990; Kim et al., 1980; and Kilburn, 1994) found higher enlistment probabilities for blacks relative to whites and lower enlistment probabilities for Hispanics relative to whites. For the senior and graduate models, our estimates of the effects of being black on enlistment probabilities were in fact smaller. This finding is consistent with results reported in another portion of this project, Orvis et al. (1996). That study also finds evidence of declining propensity for blacks relative to the 1980s. Blacks have traditionally been overrepresented in the enlisted force and have been an important source of recruits.

Does this finding imply that historic patterns of black overrepresentation will reverse? This is unlikely, because black youths in general often have other characteristics that raise the probability of enlisting, such as low family income and a large number of siblings. Given that the Hosek and Peterson papers estimate an effect of being black on the probability of enlisting that is smaller than the net effect of these other characteristics, it is likely that blacks will continue to enlist in large numbers.

Another hypothesis related to the increasing representation of minority groups in the population was the hypothesis that the growing proportion of Hispanics would have little influence on the estimate of the Hispanic variable. This was borne out in our senior estimates, where ours were not significantly different from those of Hosek and Peterson. However, our graduate estimates did differ. Hosek and Peterson estimated a negative coefficient for seniors and we estimated a positive coefficient but did not find this variable to be significant.

Due to the strong link between college attendance and family income, we also hypothesized that the rise in college attendance would make family income a more important predictor of enlistment. Our results were not consistent with this hypothesis for seniors or graduates. In both cases, our estimates were not significantly different from the estimates of Hosek and Peterson. Most of the other changes we discussed in Chapter Three were expected to influence the intercept or to have ambiguous effects on specific coefficient estimates.

Our estimates also identified some additional variables that were significant predictors of enlistment probability that Hosek and Peterson did not include in their specification. For seniors, a variable that indicates when English is not a youth's first language substantially lowered the probability of enlistment. This variable was intended to proxy for being a member of an immigrant family. The diversity of the immigrant population and the divergence in patterns of college attendance among them is great: some Asian immigrant groups have much higher college attendance rates than average, while other immigrant groups such as those from Central America have much lower college attendance rates than average. Given this diversity, our finding on the effect of English being a second language warrants more exploration. That is, it may be premature to direct recruiting efforts away from immigrant groups when some immigrant groups may yield high numbers of recruits and others yield few.

For graduates, a new variable that significantly raised enlistment probability is having a parent in the military. This finding suggests potential for recruiting through veterans' organizations or other avenues for targeting youths with currently or formerly enlisted parents. Variables indicating marijuana use yielded some ambiguous but suggestive results: youths who did not answer questions about their marijuana use, which may be correlated with use, were substantially less likely to enlist than others.

Appendix A

SUPPLEMENTARY TABLES

Table A.1

Variable Definitions

Black	Indicator for whether the individual is black and not Hispanic.
Hispanic	Indicator for whether the individual is Hispanic (can be of any race).
AFQT percentile	Percentile score of the Armed Forces Qualification Test (AFQT), estimated from the 1992 NELS test scores. See Kilburn et al. (1998). This variable is zero for those with AFQT percentile scores of 10 to 30 (CAT IV) and missing.
AFQT CAT IV	Indicator for whether information on the individual's AFQT percentile was in the 10–30 range.
AFQT missing	Indicator for whether information on the individual's AFQT percentile was missing.
Age when a high school senior	Age of the individual when a senior in high school. Entered as indicator variables for each age, with age 18 as the comparison or left-out group.
GED	Indicator for whether the individual left high school and later received a Certificate of General Education Development; applicable only to the graduate sample.
Mother's education	Years of schooling attained by the individual's mother. A continuous variable was imputed from the categorical education variables in the NELS, using the midpoint of years of education for each category. Was continuous in Hosek and Peterson data.
Mother worked	Indicator for whether the individual's mother worked outside the home. In the NELS, was for the year 1992. For the Hosek and Peterson sample, was for when the individual was age 14.

Table A.1 (continued)

Number of siblings	Number of brothers and sisters the individual has regardless of whether they still live at home.
Family income	Parental income in dollars in 1992. Values represent midpoints of income ranges that define the income category associated with the individual. The value of this variable is zero if the parental income is below $5,200 a year (the lowest income category); the "low income" indicator variable controls for these zero values. Parental income is only available for those who still lived with their parents at the time of the interview.
Lowest family income	Indicator for whether parental income was under $5,200 a year if the individual lived with his parents. We view these low income values as aberrations and chose to control for them separately to get a more accurate estimate of the effect of family income.
Family income missing	Indicator for whether information on parental income was missing if the individual lived with his parents. The values for parental income and lowest income group are zero when this variable is equal to one.
Lives at home	Indicator for whether the individual still lives with parents or guardians.

The following variables were at the county level in the Hosek and Peterson sample and were extracted for the County Statistics File produced by the Bureau of the Census. For the NELS, these data are at the state level (this was the lowest level of geographic identifier on the NELS) and are from 1990 Census files. We describe our file in detail. See Hosek and Peterson (1985, 1990) for a description of their data.

Unemployment rate	Total unemployment rate in 1990.
Percentage black	Percent of total population who are black.
Percentage Hispanic	Percent of total population who are Hispanic.
Percent of labor force female	Percent of county's total civilian labor force who are women. Figures from Bureau of the Census. Civilian labor force includes employed and unemployed civilians ages 16 and above.
Percent change in per-capita personal income	Percent increase or decrease in real per-capita personal income between 1992 and 1993 in the state.

Table A.1 (continued)

Unemployment rate and personal unemployment	Three variables were created to examine the interaction of local unemployment conditions with the individual's own unemployment situation. Interactions were made separately between the unemployment rate in 1990 and the following two individual unemployment variables: not currently working but worked in last 12 months, and number of months not employed.
Share of seniors and recent high school graduates in local market	This variable differs substantially from that used in Hosek and Peterson. They use the proportion of male youth population aged 15–24 in the MEPS (Military Entrance Processing Station) area who are high school seniors or graduated from high school in the previous year. Population figures are for 1978. We use the ratio of male high school degree graduates in the state in 1992 to the male population aged 18–24.
Worked in past year, not currently	Indicator for whether the individual had a job within the last 12 months but is not currently working.
Hourly wage	Natural log of hourly wage the individual receives at his current job. The variable is zero if the individual is not currently employed. In the senior sample, the value of this variable is zero for those with an hourly wage of less than $2.00/hour, as are all of the employment-related variables. Seniors with such extremely low hourly wages are anomalies and have been effectively removed from the estimation of the wage effect by zeroing out their values and including an indicator for such low wages.
Wage less than $2.00	Indicator for whether the individual's hourly wage was less than $2.00. This variable is applicable only to the senior sample. All the employment variables in the senior equation are zero when this variable is equal to one.
Wage missing	Indicator for whether the individual's current hourly wage is missing.
Weekly hours, if currently employed	Number of hours per week the individual works if he is working at the time of the survey. Variable is zero if not currently employed.
Weekly hours, not currently employed	Number of hours per week the individual worked at his last job if not currently working but had a job within the last 12 months. Value of variable is zero if currently employed.

Table A.1 (continued)

Prior hours missing	Indicator variable if weekly hours, not currently employed is missing.
Months on job, if currently employed	Natural log of the number of months the individual has been working on his current job. The value of this variable is zero if the individual is not currently employed.
Months not employed	Number of months since the individual's last job if he is not currently employed but had worked within the last 12 months. Value of variable is zero if currently employed.
Not employed within past year	Indicator for whether individual did not have a job during the last 12 months.
Recruiter density	Number of recruiters divided by the youth population in the state, in thousands.
Expects more education	Indicator for whether the individual's expected years of schooling exceed the number of years he has already completed.
Ever married	Indicator for whether individual is now or has ever been married.
Plans never to marry	Indicator for whether individual does not plan to get married ever.
Plans to marry in five years	Indicator for whether individual expects to be married within the next five years.
Missing marriage plans	Indicator variable equal to one when the individual is missing marriage plans information.
Has children	Indicator for whether individual has any children.
Months since last attended school	Natural log of the number of months since the individual was last enrolled in school—high school or college; applicable only to the graduate sample.
Some postsecondary schooling	Indicator of whether the individual has completed more than 12 years of schooling; applicable only to the graduate sample.

The following variables were included in the new specifications.

Parents in the military	Indicator variable equal to one if the respondent's parents were in the military in 1992. Set to zero if missing.
Missing parents in the military	Indicator variable equal to one if information on whether the respondent's parents were in the military was missing.

Table A.1 (continued)

In-state college tuition	Attendance-weighted mean in-state four-year college tuition in the state of residence in 1992. From the *Digest of Educational Statistics.*
English is second language	Indicator variable equal to one if English is not the respondent's first language.
Missing English language	Indicator variable equal to one if information on whether the respondent's first language is English was missing.
Smoked marijuana	Indicator variable equal to one if the respondent reported ever having smoked marijuana.
Missing smoked marijuana	Indicator variable equal to one if information on whether the respondent ever smoked marijuana was missing.
Arrested	Indicator variable equal to one if the respondent reported that he or a friend had ever been arrested.
Missing arrested	Indicator variable equal to one if information on whether the respondent or a friend had ever been arrested was missing.

Table A.2

Means and Standard Deviations of Variables

	Seniors		Graduates	
Variable	Mean	Standard Deviation	Mean	Standard Deviation
Black	0.1067	0.3088	0.1730	0.3783
Hispanic	0.0944	0.2924	0.1280	0.3342
Age 16 when senior	—	—	0.0040	0.0631
Age 17 when senior	0.4192	0.4935	0.3176	0.4656
Age 19+ when senior	0.1148	0.3188	0.2082	0.4061
AFQT score (31–99)	61.4393	33.7853	17.3764	26.9345
Category IV indicator	0.2177	0.4127	0.2861	0.4520
AFQT score missing	0.2344	0.4237	0.3893	0.4877
GED	—	—	0.1208	0.3259
Mother's years of schooling	13.3061	2.4110	12.4844	2.0849
Mother worked	0.8523	0.3549	0.7701	0.4208
Family income (in $ thousands)	41.7598	43.7521	28.4140	30.2063
Family income < $5,200	0.0609	0.2391	0.1170	0.3215
Family income missing	0.1798	0.3840	0.1855	0.3888
Number of siblings	1.9085	1.5661	2.0747	1.7681
Missing number of siblings	0.1176	0.3221	0.1746	0.3797
Lives at home	0.7828	0.4123	0.5480	0.4978
Missing lives at home	0.1462	0.3533	—	—
Hourly wage (natural log)	0.7411	0.8572	1.1048	0.9731
Wage < $2.25/hour	0.0217	0.1456	—	—
Hourly wage missing	0.0759	0.2648	0.0289	0.1675
Weekly hours currently employed	8.5683	11.3469	28.4866	21.3850
Months employed (natural log)	0.945	1.3192	1.8032	1.4184
Not currently employed	0.4865	0.4998	0.2735	0.4458
Weekly hours not currently employed	3.3123	8.6947	7.4237	17.1080
Missing hours not currently employed	—	—	0.0876	0.2827
Months not employed	2.69265	6.1046	2.2621	6.1789
Not employed in last 12 months	0.1757	0.3806	0.0852	0.2792
Unemployment rate in county	6.7587	1.3434	6.2505	2.2010
Share of seniors and recent graduates	0.4438	0.0696	0.4421	0.0590
Percent of population that is black	0.1217	0.0809	0.1210	0.1051
Percent of population that is Hispanic	0.0219	0.0242	0.0207	0.0243
Percent of labor force that is female	45.6189	1.2133	42.0591	12.3264
Per-capita personal income	15.0317	12.92869	1.8818	0.6131

Table A.2 (continued)

Variable	Seniors		Graduates	
	Mean	Standard Deviation	Mean	Standard Deviation
% change in per-capita personal income	0.0444	0.0183	0.0433	0.0216
Unemp rate × mos not employed	3.3316	41.7781	13.4036	39.6346
Unemp rate × not employed last 12 mos	1.2156	2.6997	0.5029	1.8023
Unemp rate × not currently employed	18.2548	3.5547	1.7278	3.0729
Recruiter density	0.0791	0.078894	0.0828	0.1046
Missing state	—	—	0.0783	0.2687
Expects more education	0.9044	0.2941	0.6185	0.4858
Missing expects more education	0.0386	0.1927	0.2793	0.4487
Plans to get married in next 5 years	0.0846	0.2783	0.0804	0.2719
Plans never to marry	0.0467	0.2111	0.0422	0.2012
Ever been married	—	—	0.1150	0.3191
Has children	—	—	0.1660	0.3722
Missing marital information	0.1846	0.3880	0.3686	0.4825
Months since school (natural log)	—	—	0.7724	1.2230
Some post-HS school	—	—	0.2471	0.4314
Parent in the military	0.0196	0.1388	0.0208	0.1426
Missing parent in military	0.0793	0.2702	0.1219	0.3272
English not first language	0.0911	0.2878	0.0993	0.2991
Missing English information	0.0357	0.1857	0.0521	0.2222
Has used marijuana	0.2561	0.4365	0.2331	0.4228
Missing marijuana info	0.1547	0.3617	0.3892	0.4876
Respondent or friend has been arrested	0.0523	0.2227	0.0570	0.2318
Missing arrest info	0.0171	0.1297	0.2596	0.4385
Mean in-state tuition	2389.2730	793.3852	2101.8300	969.1363

NOTE: A "—" indicates that the variable was not included in the model.

Table A.3

Multinomial Logit Estimates for Trivariate Model

	Attend College		Work/Other	
Variable	Coefficient Estimate	Standard Error	Coefficient Estimate	Standard Error
Black	–0.3221**	0.1608**	–0.0676	0.1542
Hispanic	–0.1721	0.2177	–0.0136	0.2120
Age 16 when senior	–0.0827	0.5013	–0.0578	0.5296
Age 17 when senior	–0.0372	0.1109	–0.0668	0.1099
Age 19+ when senior	0.1321	0.1894	0.2127	0.1765
AFQT CAT I indicator	0.1643	0.4421	–0.9056*	0.5122*
AFQT CAT II indicator	–0.1583	0.2028	–0.3872*	0.2099*
AFQT CAT IIIB indicator	–0.6488***	0.1925***	–0.3396*	0.1934*
AFQT CAT IV indicator	–0.5094**	0.2027**	0.3784*	0.2004*
AFQT CAT V indicator	–0.5071	0.3282	0.6421**	0.3111**
AFQT score missing	–0.7893***	0.1955***	–0.3024	0.1949
GED	–0.8730***	0.2787***	0.0919	0.2372
Mother's ed: less than high school	–0.5811***	0.1787***	–0.1535	0.1637
Mother's ed: some college	0.1790	0.1462	–0.1528	0.1461
Mother's ed: college degree	0.5115***	0.1793***	–0.2757	0.1845
Mother's ed: postcollegiate	0.8670***	0.2442***	0.0263	0.2531
Missing mother's education	–0.7535***	0.1565***	–0.4981***	0.1493***
Mother worked	–0.1008	0.1568	–0.3583**	0.1487**
Family income (in $ thousands)	0.0180***	0.0038***	–0.0006	0.0040
Family income < $5,200	0.4699**	0.2328**	0.4211*	0.2196*
Family income missing	0.6847***	0.1684***	0.2953*	0.1657*
Number of siblings	–0.2067***	0.0348***	–0.0811**	0.0335**
Missing number of siblings	–0.2389	0.2051	0.1547	0.1987
Unemployment rate in county	–0.1943***	0.0617***	–0.0951	0.0600
Percent of population that is black	–1.0734*	0.6501*	–1.8140***	0.6240***
Percent of population that is Hispanic	5.9498	3.8849	–0.2018	3.7663
Percent of labor force that is female	–0.0306	0.0718	–0.0308	0.0700
Per-capita personal income	1.6414***	0.5893***	1.3003**	0.5783**
% change in per-capita personal income	–1.2550	3.3173	4.4247	3.2536
Recruiter density	–2.9028***	0.6040***	–1.4783***	0.5830**
Missing state	–1.3913	3.2860	–0.4718	3.1980
Expects more education	2.0938***	0.2365***	–0.0542	0.1587

Table A.3 (continued)

Variable	Attend College		Work/Other	
	Coefficient Estimate	Standard Error	Coefficient Estimate	Standard Error
Missing expects more education	2.1999***	0.4914***	0.6375	0.4430
Plans to get married in next 5 years	–0.8987***	0.1936***	–0.1421	0.1812
Plans never to marry	–0.8217***	0.2195***	–0.2573	0.2120
Ever been married	–2.1319***	0.2369***	–0.6316***	0.1836***
Has children	–0.5494**	0.2335**	0.4430**	0.1921**
Missing marital information	0.1573	0.2100	0.2489	0.2005
Parent in the military	–2.0422***	0.2758***	–1.2821***	0.2484***
Missing parent in military	–0.3771*	0.2259*	–0.0167	0.2156
English not first language	1.1139***	0.2415***	0.3429	0.2382
Missing English language info	0.9854***	0.3716***	0.1188	0.3626
Uses marijuana	–0.1252	0.1253	0.3760**	0.1220***
Missing marijuana use	–0.0310	0.2228	0.5402**	0.2124**
R or friend has been arrested	–0.4800**	0.2283**	–0.0071	0.2128
Missing arrest info	–0.4214	0.4726	0.8541*	0.4457*
Average in-state tuition	–0.0001	0.0001	–0.0002	0.0001
Constant	2.2929	3.2920	3.4367	3.2044
Log-likelihood	–5204.96			
Chi-square	3790.71			
Pseudo R-square	0.2669			
Number of observations	8,009			

NOTE: Significance levels: *** 0.01 level, ** 0.05 level, * 0.10 level.

Appendix B

TESTING FOR DIFFERENCES FROM HOSEK AND PETERSON RESULTS

This appendix provides more details on the tests used to determine whether the NELS coefficient estimates and regressor means are different from those in the Hosek and Peterson (1990) paper.

First we test for differences in estimated coefficients. Let the estimated coefficient for any characteristic in the Hosek and Peterson paper be $\hat{\beta}_1$ and the coefficient for that same characteristic in our paper be $\hat{\beta}_2$. Our null hypothesis is that there is no difference in the coefficients: $H_0: \hat{\beta}_1 - \hat{\beta}_2 = 0$. The test statistic for this problem is

$$t = \frac{\hat{\beta}_1 - \hat{\beta}_2}{\sqrt{\mathrm{var}\left(\hat{\beta}_1\right) + \mathrm{var}\left(\hat{\beta}_2\right)}},$$

which approximates a t-test. Both the Hosek and Peterson results and our results report the square root of the variance of the coefficient estimates, or the standard errors, S_1 and S_2. Recall that Hosek and Peterson use a choice-based sample that contains more enlistees than nonenlistees. Using a choice-based sample produces estimates that are inconsistent and inefficient without using some type of correction. Hosek and Peterson make the appropriate corrections to account for the fact that they are not using a random sample (see Hosek and Peterson, 1985, Appendix C), and we use the corrected standard errors they report. The test statistic that we compute from the regression output is

$$t = \frac{\hat{\beta}_1 - \hat{\beta}_2}{\sqrt{(S_1)^2 + (S_2)^2}}$$

Second, we test whether there are significant differences in the means of the regressors Hosek and Peterson use from a 1980 sample and the regressors we use from a 1992 sample. Let $\bar{x}_1$ be the mean of a characteristic in the Hosek and Peterson sample and $\bar{x}_2$ be the mean of the same characteristic in our sample. The null hypothesis is that there is no difference between the means of the same variable in the two samples: $H_0: \bar{x}_1 - \bar{x}_2 = 0$. The test statistic is

$$t = \frac{\bar{x}_1 - \bar{x}_2}{\sqrt{\text{var}(\bar{x}_1) + \text{var}(\bar{x}_2)}},$$

which has a student's *t* distribution. The variance of the mean of the characteristic is

$$\text{var}(\bar{x}_j) = \frac{s_j^2}{n_j}. \qquad (1)$$

Our paper reports the mean and standard deviation of the values of the observations of each characteristic, $\bar{x}_2$ and s_2, respectively. So when computing the test statistic, we can just substitute in the values for $\bar{x}_2$ and s_2^2 / n_2 from our sample.

Hosek and Peterson (1990) do not report the overall sample mean, $\bar{x}_1$, but rather the mean for the enlistee sample, $\bar{x}_m$, and the mean for the nonenlistee sample, $\bar{x}_c$. Using the population percentage of seniors or graduates who enlisted, w_m, we computed the overall sample mean for the Hosek and Peterson sample as

$$\bar{x}_1 = w_m \bar{x}_m + w_c \bar{x}_c.$$

Similarly, rather than reporting the sample standard deviation, s_1, Hosek and Peterson report a standard deviation for the enlistee sample, s_m, and for the nonenlistee sample, s_c. Hence, we calculate the $\mathrm{var}(\bar{x}_1)$ as

$$\mathrm{var}(\bar{x}_1) = \mathrm{var}(w_m \bar{x}_m + w_c \bar{x}_c).$$

Using rules for the variance of a random variable times a constant, we can rewrite this as

$$\mathrm{var}(\bar{x}_1) = w_m^2 \,\mathrm{var}(\bar{x}_m) + w_c^2 \,\mathrm{var}(\bar{x}_c).$$

Using (1) again, we can write the variance of the mean from the enlistee sample and the variance of the mean of the nonenlistee sample as

$$\mathrm{var}(\bar{x}_m) = \frac{s_m^2}{n_m} \text{ and } \mathrm{var}(\bar{x}_c) = \frac{s_c^2}{n_c}.$$

Then our expression for the total sample variance becomes:

$$\mathrm{var}(\bar{x}_1) = w_m^2 \frac{s_m^2}{n_m} + w_c^2 \frac{s_c^2}{n_c},$$

which is a function of statistics Hosek and Peterson report in their paper: the fraction of enlistees and nonenlistees, w_m and w_c, the sample characteristic standard deviations, s_m^2 and s_c^2, and the enlistee and nonenlistee sample sizes, n_m and n_c.

We therefore compute the test statistic as follows, using information reported in the two papers:

$$t = \frac{\bar{x}_1 - \bar{x}_2}{\sqrt{\left(w_m^2 \frac{s_m^2}{n_m} + w_c^2 \frac{s_c^2}{n_c} \right) + \frac{s_2^2}{n_2}}}.$$

REFERENCES

Asch, Beth J., *Navy Recruiter Productivity and the Freeman Plan,* Santa Monica, CA: RAND, R-3713-FMP, June 1990.

Asch, Beth J., and James N. Dertouzos, *Educational Benefits Versus Enlistment Bonuses: A Comparison of Recruiting Options,* Santa Monica, CA: RAND, MR-302-OSD, 1994.

Asch, Beth J., and Lynn A. Karoly, *The Role of the Job Counselor in the Military Enlistment Process,* Santa Monica, CA: RAND, MR-315-P&R, 1993.

Asch, Beth J., and Bruce R. Orvis, *Recent Recruiting Trends and Their Implications: Preliminary Analysis and Recommendations,* Santa Monica, CA: RAND, MR-549-A/OSD, 1994.

Bean, Frank D., and Marta Tienda, *The Hispanic Population of the United States,* New York: Russell Sage Foundation, 1987.

Becker, Gary S., *Human Capital: A Theoretical and Empirical Analysis,* with Special Reference to Education Second Edition, Chicago and London: The University of Chicago Press, 1975.

Binkin, Martin, and Mark Eitelberg, "Women and Minorities in the All-Volunteer Force," in William Bowman, Roger Little, and Thomas Sicilia (eds.), *The All-Volunteer Force After a Decade,* Washington, D.C.: Pergamon-Brassey's, 1986.

Binkin, Martin, Mark J. Eitelberg, Alvin J. Schneider, and Marvin M. Smith, *Blacks and the Military,* Washington, D.C.: The Brookings Institution, 1982.

Dertouzos, James N., *Recruiter Incentives and Enlistment Supply,* Santa Monica, CA: RAND, R-3065-MIL, May 1985.

Dertouzos, James N., and J. Michael Polich, *Recruiting Effects of Army Advertising,* Santa Monica, CA: RAND, R-3577-FMP, January 1989.

Eitelberg, Mark J., *Manpower for Military Occupations,* Washington, D.C.: Office of the Assistant Secretary of Defense (Force Management and Personnel), April 1988.

Fernandez, Richard L., *Enlistment Effects and Policy Implications of the Educational Assistance Test Program,* Santa Monica, CA: RAND, R-2935-MRAL, September 1982.

Gorman, Linda, and George W. Thomas, "General Intellectual Achievement, Enlistment Intentions, and Racial Representativeness of the U.S. Military," *Armed Forces and Society,* Vol. 19, No. 4, Summer 1993, pp. 611–624.

Greene, William H., *Econometric Analysis,* 2d ed., New York: Macmillan, 1990.

Harrell, Margaret C., and Laura L. Miller, *New Opportunities for Military Women: Effects Upon Readiness, Cohesion and Morale,* Santa Monica, CA: RAND, MR-896-OSD, 1997.

Hosek, James R., and Christine E. Peterson, *Enlistment Decisions of Young Men,* Santa Monica, CA: RAND, R-3238-MIL, July 1985.

Hosek, James, and Christine Peterson, *Serving Her Country: An Analysis of Women's Enlistment,* Santa Monica, CA: RAND, R-3853-FMP, 1990.

Hosek, James, and Mark Totten, *Does Perstempo Hurt Reenlistment? The Effect of Long or Hostile Perstempo on Reenlistment,* Santa Monica, CA: RAND, MR-990-OSD, 1999.

Hosek, James R., Christine E. Peterson, and Joanna Zorn Heilbrunn, *Military Pay Gaps and Caps,* Santa Monica, CA: RAND, MR-368-P&R, 1994.

Kane, Thomas J., "College Entry by Blacks Since 1970: The Role of College Costs, Family Background, and the Returns to Education," *Journal of Political Economy,* Vol. 102, No. 5, 1994, pp. 878–911.

Kilburn, M. Rebecca, "Minority Representation in the U.S. Military," unpublished dissertation, Department of Economics, University of Chicago, 1994.

Kilburn, M. Rebecca, Lawrence M. Hanser, and Jacob A. Klerman, *Estimating AFQT Scores for NELS Respondents,* Santa Monica, CA: RAND, MR-818-OSD/A, 1998.

Kim, Choongsoo, Gilbert Nestel, Robert L. Phillips, and Michael E. Borus, "The All-Volunteer Force: An Analysis of Youth Participation, Attrition, and Reenlistment," *National Longitudinal Survey of Youth Labor Market Experience Military Studies,* Center for Human Resource Research, The Ohio State University, May 1980.

Klerman, Jacob Alex, and Lynn A. Karoly, *Trends and Future Directions in Youth Labor Markets,* Santa Monica, CA: RAND, RP-310, 1994.

Levy, Frank, and Richard J. Murnane, "Earnings Level and Earnings Inequality: A Review of Recent Trends and Proposed Explanations," *Journal of Economic Literature,* Vol. 30, September 1992, pp. 133–138.

Manski, Charles F., and David A. Wise, *College Choice in America,* Cambridge, MA: Harvard University Press, 1983.

McFadden, David. "Econometric Analysis of Qualitative Response Models," in Zvi Griliches and M. Intrilligator (eds.), *Handbook of Econometrics,* Vol. 2, Amsterdam: North Holland, 1983.

Murray, Michael P., and Laurie L. McDonald, *Recent Recruiting Trends and Their Implications for Models of Enlistment Supply,* Santa Monica, CA: RAND, MR-847-OSD/A, 1998.

Office of the Assistant Secretary of Defense (Personnel and Readiness), *Population Representation in the Military Services, Fiscal Year 1992,* Washington, D.C., October 1993.

Office of the Assistant Secretary of Defense (Force Management Policy), *Population Representation in the Military Services, Fiscal Year 1994,* Washington, D.C., December 1995.

Office of the Assistant Secretary of Defense (Force Management Policy), *Population Representation in the Military Services, Fiscal Year 1995,* Washington, D.C., December 1996.

Office of the Secretary of Defense, *Military Compensation Background Papers, Compensation Elements and Related Manpower Cost Items, Their Purposes and Legislative Backgrounds,* 4th ed., Washington, D.C., November 1991.

Oken, Carole, and Beth J. Asch, *Encouraging Recruiter Achievement: A Recent History of Recruiter Incentive Programs,* Santa Monica, CA: RAND, MR-845-OSD/A, 1997.

Orvis, Bruce R., and Martin T. Gahart, *Relationship of Enlistment Intention and Market Survey Information to Enlistment in Active Duty Military Service,* Santa Monica, CA: RAND, N-2292-MIL, June 1985.

Orvis, Bruce R., and Martin T. Gahart, *Quality-Based Analysis Capability for National Youth Surveys: Development, Application, and Implications for Policy,* Santa Monica, CA: RAND, R-3675-FMP, March 1989.

Orvis, Bruce R., and Martin T. Gahart, *Enlistment Among Applicants for Military Service: Determinants and Incentives,* Santa Monica, CA: RAND, R-3359-FMP, January 1990.

Orvis, Bruce R., Narayan Sastry, and Laurie L. McDonald, *Military Recruiting Outlook: Recent Trends in Enlistment Propensity and Conversion of Potential Enlisted Supply,* Santa Monica, CA: RAND, MR-677-A/OSD, 1996.

Phillips, Robert L., Paul J. Andrisani, Thomas N. Daymont, and Curtis L. Gilroy, "The Economic Returns to Military Service: Race-Ethnic Differences," *Social Science Quarterly,* Vol. 73, No. 2, 1992, pp. 340–359.

Reville, Robert T., "Two Essays on Intergenerational Earnings and Wage Mobility," Ph.D. dissertation, Brown University, May 1996.

Segal, David R., *Recruiting for Uncle Sam,* Lawrence, KS: University Press of Kansas, 1989.

Segal, David, Jerald Bachman, and Faye Dowdell, "Military Service for Female and Black Youth: A Perceived Mobility Opportunity," *Youth and Society,* Vol. 10, 1978, pp. 127–134.

Schoeni, Robert F., Kevin F. McCarthy, and Georges Vernez, *The Mixed Economic Progress of Immigrants,* Santa Monica, CA: RAND, MR-763-IF/FF, 1996.

Thomas, George W., "Military Parental Effects and Career Orientation Under the AVF: Enlisted Personnel," *Armed Forces and Society,* Vol. 10, No. 2, Winter 1984, pp. 293–310.

U.S. Bureau of the Census, *Statistical Abstract of the United States,* (114th ed.), Washington, D.C., various years.

Willis, Robert, and Sherwin Rosen. "Education and Self-Selection," *Journal of Political Economy,* Vol. 87, No. 5 (Part 2), 1979, pp. S7–S36.